P9-DGY-581

Bee Cave Public Library
4000 Galleria Parkway
Bee Cave, Texas 78738

THE MATH
OF
LIFE & DEATH

---

7 MATHEMATICAL PRINCIPLES THAT SHAPE OUR LIVES

KIT YATES

SCRIBNER

NEW YORK LONDON TORONTO SYDNEY NEW DELHI

Scribner
An Imprint of Simon & Schuster, Inc.
1230 Avenue of the Americas
New York, NY 10020

Copyright © 2019 by Kit Yates
Originally published in Great Britain in 2019 by Quercus
as *The Maths of Life and Death*
Published under license from Quercus Editions Limited

All rights reserved, including the right to reproduce this book or
portions thereof in any form whatsoever. For information, address
Scribner Subsidiary Rights Department,
1230 Avenue of the Americas, New York, NY 10020.

First Scribner hardcover edition January 2020

SCRIBNER and design are registered trademarks of The Gale Group, Inc.,
used under license by Simon & Schuster, Inc., the publisher of this work.

For information about special discounts for bulk purchases,
please contact Simon & Schuster Special Sales at 1-866-506-1949
or business@simonandschuster.com.

The Simon & Schuster Speakers Bureau can bring authors to
your live event. For more information or to book an event, contact
the Simon & Schuster Speakers Bureau at 1-866-248-3049 or visit
our website at www.simonspeakers.com.

Interior design by Kyle Kabel

Illustrations by Amber Anderson

Manufactured in the United States of America

1   3   5   7   9   10   8   6   4   2

Library of Congress Cataloging-in-Publication Data
Names: Yates, Kit, author.
Title: The math of life and death : 7 mathematical principles
that shape our lives / Kit Yates.
Description: First Scribner hardcover edition. | New York : Scribner, 2020. |
Includes bibliographical references and index. |
Identifiers: LCCN 2019024831 (print) | LCCN 2019024832 (ebook) |
ISBN 9781982111878 (hardcover) | ISBN 9781982111885 (paperback) |
ISBN 9781982111892 (ebook)
Subjects: LCSH: Mathematics—Popular works.
Classification: LCC QA93 .Y38 2020 (print) | LCC QA93 (ebook) | DDC 510—dc23
LC record available at https://lccn.loc.gov/2019024831
LC ebook record available at https://lccn.loc.gov/2019024832

ISBN 978-1-9821-1187-8
ISBN 978-1-9821-1189-2 (ebook)

Note to Readers: Some names have been changed.

For my parents, Tim, Nancy, and Mary, who taught me how to read, and my sister, Lucy, who taught me how to write.

CONTENTS

ALMOST EVERYTHING

M y four-year-old son loves playing out in the garden. His favorite activity is digging up and inspecting creepy crawlies, especially snails. If he is patient enough, after the initial shock of being uprooted, they will emerge cautiously from the safety of their shells and start to glide over his little hands, leaving viscid trails of mucus. Eventually, when he tires of them, he will discard them, somewhat callously, in the compost heap or on the woodpile behind the shed.

Late last September, after a particularly busy session in which he had unearthed and disposed of five or six large specimens, he came to me as I was sawing up wood for the fire and asked, "Daddy, how many snails is [*sic*] there in the garden?" A deceptively simple question for which I had no good answer. It could have been one hundred or it could have been one thousand. He would not have comprehended the difference. Nevertheless, his question piqued an interest in me. How could we figure this out together?

We decided to conduct an experiment. The next weekend, on Saturday morning, we went out to collect snails. After ten minutes, we had a total of 23 of the gastropods. I took a Sharpie from my back pocket and placed a subtle cross on the back of each. Once they were all marked up, we tipped up the bucket and released the snails back into the garden.

A week later we went back out for another round. This time, our ten-minute scavenge brought us just 18 snails. When we inspected them closely, we found that 3 of them had the cross on their shells, while the other 15 were unblemished. This was all the information we needed to make the calculation.

The idea is as follows: The number of snails we captured on the first day, 23, is a given proportion of the total population of the garden, which we want to get a handle on. If we can work out this proportion, then we can scale up from the number of snails we caught to find the total population of the garden. So we use a second sample (the one we took the following Saturday). The proportion of marked individuals in this sample, 3/18, should be representative of the proportion of marked individuals in the garden as a whole. When we simplify this proportion, we find that the marked snails make up one in every six individuals in the population at large (you can see this illustrated in figure 1). Thus we scale up the number of marked individuals caught on the first day, 23, by a factor of six to find an estimate for the total number of snails in the garden, which is 138.

After finishing this mental calculation I turned to my son, who had been "looking after" the snails we had collected. What did he make of it when I told him that we had roughly 138 snails living in our garden? "Daddy"—he looked down at the fragments of shell still clinging to his fingers—"I made it dead." Make that 137.

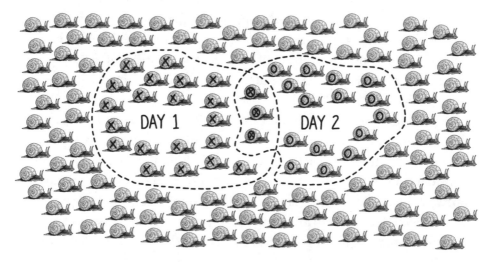

FIGURE 1: The ratio of snails recaptured (marked ⊗) to the total captured (marked O) on day 2 is 3:18, which should be the same as the ratio of snails captured on day 1 (marked ×) to all snails in the garden, 23:138.

This simple mathematical method, known as capture-recapture, comes from ecology, where it is used to estimate animal population sizes. You can use the technique yourself: take two independent samples and compare the overlap between them. Perhaps you want to estimate the number of raffle tickets that were sold at the local fair or to estimate the attendance at a football match using ticket stubs rather than having to do an arduous head count.

Capture-recapture is used in serious scientific projects as well. It can, for example, give vital information on the fluctuating numbers of an endangered species. By providing an estimate of the number of fish in a lake, it might allow fisheries to determine how many permits to issue. Such is the effectiveness of the technique that its use has evolved beyond ecology to provide accurate estimates on everything from the number of drug addicts in a population to the number of war dead in Kosovo. This is the pragmatic power that simple mathematical ideas can wield. These are the sorts of concepts that we will explore throughout this book and that I use routinely in my day job as a mathematical biologist.

When I tell people I am a mathematical biologist, I usually get a polite nodding of the head accompanied by an awkward silence, as if I were about to test them on their recall of the quadratic formula or Pythagoras's theorem. More than simply being daunted, people struggle to understand how a subject such as math, which they perceive as being abstract, pure, and ethereal, can have anything to do with a subject such as biology, which is typically thought of as being practical, messy, and pragmatic. This artificial dichotomy is often first encountered at school: If you liked science but you weren't so hot on algebra, then you were pushed down the life sciences route. If, like me, you enjoyed science but you weren't into cutting up dead things (I fainted once, at the start of a dissection class, when I walked into the lab and saw a fish head sitting at my bench space), then you were guided toward the physical sciences. Never the twain shall meet.

This happened to me. I dropped biology at sixth form and took A levels in math, further math, physics, and chemistry. When it came to university, I had to further streamline my subjects and felt sad that I had to leave biology behind forever: a subject that I thought had incredible power to change lives for the better. I was hugely excited about the opportunity to plunge myself into the world of mathematics, but I couldn't help worrying that I was taking on a subject that seemed to have few practical applications. I couldn't have been more wrong.

While I plodded through the pure math we were taught at university, memorizing the proof of the intermediate value theorem or the definition of a vector space, I lived for the applied-math courses. I listened to lecturers as they demonstrated the math that engineers use to build bridges so that they don't resonate and collapse in the wind, or to design wings that ensure planes don't fall out of the sky. I learned the quantum mechanics that physicists use to understand the strange goings-on at subatomic scales, and the theory of special relativity, which explores the strange consequences of the invariance of the speed of light. I took courses explaining the ways in which we use mathematics in chemistry, in finance, and in economics. I read about how we use mathematics in sports to enhance the performance of our top athletes, and how we use mathematics in the movies to create computer-generated images of scenes that couldn't exist in reality. In short, I learned that mathematics can be used to describe almost everything.

In the third year of my degree I was fortunate enough to take a course in mathematical biology. The lecturer was Philip Maini, an engaging Northern Irish professor in his forties. Not only was he the preeminent figure in his field (he would later be elected to the Fellowship of the Royal Society), but he clearly loved his subject, and his enthusiasm spread to the students in his lecture theater.

More than just mathematical biology, Philip taught me that mathematicians are human beings with feelings, not the one-dimensional automatons that they are often portrayed to be. A mathematician is more than just, as the Hungarian probabilist Alfréd Rényi once put it, "a machine for turning coffee into theorems." As I sat in Philip's office awaiting the start of the interview for a PhD place, I saw, framed on the walls, the numerous rejection letters he had received from the Premier League clubs to whom he had jokingly applied for vacant managerial positions. We ended up talking more about football than we did about math.

Crucially at this point in my academic studies, Philip helped me to become fully reacquainted with biology. During my PhD under his supervision, I worked on everything, from understanding the way locusts swarm and how to stop them, to predicting the complex choreography that is the development of the mammalian embryo and the devastating consequences when the steps get out of sync. I built models explaining how birds' eggs get their beautiful pigmentation patterns and wrote algorithms to track the movement of free-swimming bacteria. I simulated parasites evading our immune systems and modeled the way in which deadly diseases spread through a population. The work I started during my PhD has been the bedrock for the rest of my career. I still work on these fascinating areas of biology, and others, with PhD students of my own, in my current position as an associate professor (senior lecturer) in applied mathematics at the University of Bath.

As an applied mathematician, I see mathematics as, first and foremost, a practical tool to make sense of our complex world. Mathematical modeling can give us an advantage in everyday situations, and it doesn't have to involve hundreds of tedious equations or lines of computer code to do so. Mathematics, at its most fundamental, is pattern. Every time you look at the world you are building your own model of the patterns you observe. If you spot a motif in the fractal branches of a tree, or in the multifold symmetry of a snowflake, then you are seeing math. When you tap your foot in time to a piece of music, or when your voice reverberates and resonates as you sing in the shower, you are hearing math. If you bend a shot into the back of the net or catch a baseball on its parabolic trajectory, then you are doing math. With every new experience, every piece of sensory information, the models you've made of your environment are refined, reconfigured, and rendered ever more detailed and complex. Building mathematical models designed to capture our intricate reality is the best way we have of making sense of the rules that govern the world around us.

I believe that the simplest, most important models are stories and analogies. The key to exemplifying the influence of the unseen undercurrent of math is to demonstrate its effects on people's lives: from the extraordinary to the everyday. When viewing through the correct lens, we can start to tease out the hidden mathematical rules that underlie our common experiences.

The seven chapters of this book explore the true stories of life-changing events in which the application (or misapplication) of mathematics has played a critical role: patients crippled by faulty genes and entrepreneurs bankrupt by faulty algorithms; innocent victims of miscarriages of justice and the unwitting victims of software glitches. We follow stories of investors who have lost fortunes and parents who have lost children, all because of mathematical misunderstanding. We wrestle with ethical dilemmas from screening to statistical subterfuge and examine pertinent societal issues such as political referenda, disease prevention, criminal justice, and artificial intelligence. In this book we will see that mathematics has something profound or significant to say on all of these subjects, and more.

Rather than just pointing out the places in which math might crop up, throughout these pages I will arm you with simple mathematical rules and tools that can help you in your everyday life: from getting the best seat on the train, to keeping your head when you get an unexpected test result from the doctor. I will suggest simple ways to avoid making numerical mistakes, and we will get our hands dirty with newsprint when untangling the figures behind the headlines. We will also get up close and personal with the math behind consumer genetics and observe math in action as we highlight the steps we can take to help halt the spread of a deadly disease.

I hope you'll have worked out by now that this is not a math book. Nor is it a book for mathematicians. You will not find a single equation in these pages. The point of the book is not to bring back memories of the school mathematics lessons you might have given up years ago. Quite the opposite. If you've ever been disenfranchised and made to feel that you can't take part in mathematics or aren't good at it, consider this book an emancipation.

I genuinely believe that math is for everyone and that we can all appreciate the beautiful mathematics at the heart of the complicated phenomena we experience daily. As we will see in the following chapters, math is the false alarms that play on our minds and the false confidence that helps us sleep at night; the stories pushed at us on social media and the memes that spread through it. Math is the loopholes in the law and the needle that closes them; the technology that saves lives and the mistakes that put them at risk; the outbreak of a deadly disease and the strategies to control it. It is the best hope we have of answering the most fundamental questions about the enigmas of the cosmos and the mysteries of our own species. It leads us on the myriad paths of our lives and lies in wait, just beyond the veil, to stare back at us as we draw our final breaths.

CHAPTER 1

THINKING EXPONENTIALLY

The Sobering Limits of Power

D arren Caddick is a driving instructor from a small town in South Wales. In 2009, he was approached by a friend with a lucrative offer. By contributing just £3,000 to a local investment syndicate and recruiting two more people to do the same, he would see a return of £23,000 in just a couple of weeks. Initially, thinking it was too good to be true, Caddick resisted the temptation. Eventually, though, his friends convinced him that "nobody would lose, because the scheme would just keep going and going and going," so he decided to throw in his lot.

Unwittingly, Caddick had found himself at the bottom of a pyramid scheme that couldn't "just keep going." Initiated in 2008, the Give and Take scheme ran out of new investors and collapsed in less than a year, but not before sucking in £21 million from over ten thousand investors across the UK, 90 percent of whom lost their £3,000 stake. Investment schemes that rely on investors recruiting multiple others to realize their payout are doomed to failure. The number of new investors needed at each level increases in proportion to the number of people in the scheme. After fifteen rounds of recruitment, there would be over ten thousand people in a pyramid scheme of this sort. Although that sounds like a large number, it was easily achieved by Give and Take. Fifteen rounds further on, however, and one in every seven people on the planet would need to invest to keep the scheme going. This rapid growth phenomenon, which led to an inevitable lack of new recruits and the eventual collapse of the scheme, is known as exponential growth.

No Use Crying over Spoiled Milk

Something grows exponentially when it increases in proportion to its current size. Imagine, when you open your pint of milk in the morning, a single cell of the bacteria *Streptococcus faecalis* finds its way into the bottle before you put the lid back on. *Strep f.* is one of the bacteria responsible for the souring and curdling of milk, but one cell is no big deal, right? Maybe it's more worrying when you find out that, in milk, *Strep f.* cells can divide to produce two daughter cells every hour. At each generation, the number of cells increases in proportion to the current number of cells, so their numbers grow exponentially.

The curve that describes how an exponentially growing quantity increases is reminiscent of a quarter-pipe ramp used by skaters, skateboarders, and BMXers. Initially, the gradient of the ramp is very low—the curve is extremely shallow and gains height only gradually (as you can see from the first curve in figure 2). After two hours four *Strep f.* cells are in your milk, and after four hours there are still only sixteen, which doesn't sound like too much of a problem. As with the quarter pipe, though, the height of the exponential curve and its steepness rapidly increase. Quantities that grow exponentially might appear to grow slowly at first, but they can take off quickly in a way that seems unexpected. If you leave your milk out on the side for forty-eight hours, and the exponential increase of *Strep f.* cells continues, when you pour it on your cereal again, there could be almost a thousand trillion cells in the bottle—enough to make your blood curdle, let alone the milk. At this point the cells would outnumber the people on our planet forty thousand to one. Exponential curves are sometimes referred to as J-shaped as they almost mimic the letter J's steep curve. As the bacteria use up the nutrients in the milk and change its pH, the growth conditions deteriorate, and the exponential increase is only sustained for a relatively short time. Indeed, in almost every real-world scenario, long-term exponential growth is unsustainable, and in many cases pathological, as the subject of the growth uses up resources in an unviable manner. Sustained exponential growth of cells in the body, for example, is a typical hallmark of cancer.

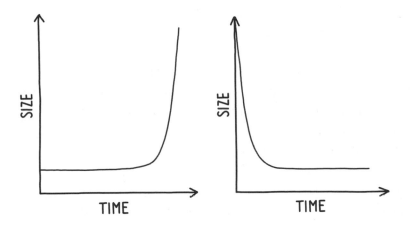

FIGURE 2: J-shaped exponential growth (*left*) and decay (*right*) curves.

Another example of an exponential curve is a free-fall waterslide, so called because the slide is initially so steep that the rider feels the sensation of free fall. This time, as we travel down the slide, we are surfing an exponential *decay* curve, rather than a *growth* curve (you can see an example of such a graph in the second image of figure 2). Exponential decay occurs when a quantity *decreases* in proportion to its current size. Imagine opening a huge bag of M&M's, pouring them out onto the table, and eating all the sweets that land with the M-side facing upward. Put the rest back in the bag for tomorrow. The next day give the bag a shake and pour out the M&M's. Again, eat the M-up sweets and put the rest back in the bag. Each time you pour the sweets out of the bag, you get to eat roughly half of those that remain, irrespective of the number you start with. The number of sweets decreases in proportion to the number left in the bag, leading to exponential decay in the number of sweets. In the same way, the exponential waterslide starts high up and almost vertical, so that the height of the rider decreases rapidly; when we have large numbers of sweets, the number we get to eat is also large. But the curve ever-so-gradually gets less and less steep until it is almost horizontal toward the end of the slide; the fewer sweets we have left, the fewer we get to eat each day. Although an individual sweet landing M-up or M-down is random and unforeseeable, the predictable water-slide curve of exponential decay emerges in the number of sweets we have left over time.

Throughout this chapter we will uncover the hidden connections between exponential behavior and everyday phenomena: the spread of a disease through a population or a meme through the internet; the rapid growth of an embryo or the all-too-slow growth of money in our bank accounts; the way in which we perceive time and even the explosion of a nuclear bomb. As we progress, we will carefully unearth the full tragedy of the Give and Take pyramid scheme. The stories of the people whose money was sucked in and swallowed serve to illustrate just how important it is to be able to think exponentially, which will in turn help us anticipate the sometimes surprising pace of change in the modern world.

A Matter of Great Interest

On the all-too-rare occasions when I get to make a deposit into my bank account, I take solace in that no matter how little I have in there, it is always growing exponentially. Indeed, a bank account is one place where there are genuinely no limits on exponential growth, at least on paper. Provided that the interest is compounded (i.e., interest is added to our initial amount and earns interest itself), then the total amount in the account increases in proportion to its current size—the hallmark of exponential growth. As Benjamin Franklin put it, "Money makes money, and the money that money makes, makes more money." If you could wait long enough, even the smallest investment would become a fortune. But don't go and lock up your rainy-day fund just yet. If you invested $100 at 1 percent per year, it would take you over nine hundred years to become a millionaire. Although exponential growth is often associated with rapid increases, if the rate of growth and the initial investment are small, exponential growth can feel very slow indeed.

The flip side of this is that, because you are charged a fixed rate of interest on the outstanding amount (often at a large rate), debts on credit cards can also grow exponentially. As with mortgages, the earlier you pay your credit cards off and the more you pay early on, the less you end up paying overall, as exponential growth never gets a chance to take off.

Paying off mortgages and sorting out other debts was one of the main reasons given by victims of the Give and Take scheme for getting involved in the first place. The temptation of quick and easy money to reduce financial pressures was too much for many to resist, despite the nagging suspicion that something wasn't quite right. As Caddick admits, "The old adage of 'If something looks too good to be true, then it probably is,' is really, really true here."

The scheme's initiators, pensioners Laura Fox and Carol Chalmers, had been friends since their days at a Catholic convent school. The pair, both pillars of their local community—one vice president of her local Rotary Club, the other a well-respected grandmother—cunningly designed the investment scheme. Give and Take was cleverly designed to ensnare potential investors, while hiding the pitfalls. Unlike the traditional two-level pyramid scheme, in which the person at the top of the chain takes money directly from the investors they have recruited, Give and Take operated as a four-level "airplane" scheme. In an airplane scheme, the person at the top of the chain is known as the "pilot." The pilot recruits two "copilots," who each recruit two "crew members," who finally each recruit two "passengers." In Fox and Chalmers's scheme, once the hierarchy of fifteen people was complete, the eight passengers paid their £3,000 to the organizers, who passed a huge £23,000 payout to the initial investor with £1,000 skimmed off the top. Part of this money was donated to charity, with letters of thanks from the likes of the NSPCC (National Society for the Prevention of Cruelty to Children) adding legitimacy to the scheme. Part was kept by the organizers to ensure the continued smooth operation of the scheme.

Having received a payout, the pilot then drops out of the scheme, and the two copilots are promoted to pilot, awaiting the recruitment of eight new passengers at the bottom of their trees. Airplane schemes are particularly seductive for investors, as new participants need only recruit two other people to multiply their investment by a factor of eight (although these two are required to recruit two more and so on). Other, flatter, schemes require far more recruitment effort per individual for the same returns. The steep four-level structure of Give and Take meant that crew members never took money directly from the passengers they recruited. Since new recruits are likely to be friends and relatives of the crew members, this ensures that money never travels directly between close acquaintances. This separation of the passengers from the pilots, whose payouts they fund, renders recruitment easier and reprisals less likely, making for a more attractive investment opportunity and thus facilitating the recruitment of thousands of investors to the scheme.

In the same way, many investors in the Give and Take pyramid scheme were given the confidence to invest by stories of successful payouts that had previously been made, and in some cases by even witnessing these payouts firsthand. The scheme's organizers, Fox and Chalmers, hosted lavish private parties at the Somerset hotel owned by Chalmers. Flyers handed out at the parties included pictures of the scheme's members, sprawled on cash-covered beds or waving fists of fifties at the camera. To each of these parties the organizers also invited some of the scheme's "brides"—those people (mainly women) who had made it to the position of pilot of their pyramid cell and were due to receive their payouts. The brides would be asked a series of four simple questions—such as "What part of Pinocchio grows when he lies?"—in front of an audience of two hundred to three hundred potential investors.

This "quiz" aspect of the scheme was supposed to exploit a loophole in the law, which Fox and Chalmers believed allowed for such investments if an element of "skill" was involved. In mobile-phone footage of one such event, Fox can be heard shouting, "We are gambling in our own homes and that's what makes it legal." She was wrong. Miles Bennet, the lawyer prosecuting the case, explained, "The quiz was so easy that there were never any people in the payout position who didn't get their money. They could even get a friend or a committee member to help with the questions, and the committee knew what the answers were!"

This didn't stop Fox and Chalmers from using these prize-giving parties as inoculants in their low-tech viral-marketing campaign. Upon seeing the brides presented with their £23,000 checks, many of the invited guests would invest and encourage their friends and family to do the same, forming the pyramid beneath them. Providing each new investor passed the baton to two or more others, the scheme would continue indefinitely. When Fox and Chalmers started the scheme, back in the spring of 2008, they were the only two pilots. By recruiting friends to invest and indeed help organize the scheme, the pair quickly brought four more people on board. These four recruited eight more and then sixteen and so on. This exponential doubling of the number of new recruits in the scheme closely mimics the doubling of the number of cells in a growing embryo.

The Exponential Embryo

When my wife was pregnant with our first child, we were obsessed, like many first-time parents-to-be, by trying to find out what was going on inside my wife's midriff. We borrowed an ultrasound heart monitor to listen to our baby's heartbeat; we signed up for clinical trials to get extra scans; and we read website after website describing what was going on with our daughter as she grew and continued to make my wife sick every day. Among our "favorites" were the "How big is your baby?"–type websites, which compare, for each week of gestation, the size of an unborn baby to a common fruit, vegetable, or other appropriately sized foodstuff. They give substance to prospective parents' unborn fetuses with epigrams such as "Weighing about one and a half ounces and measuring about three and a half inches, your little angel is roughly the size of a lemon" or "Your precious little turnip now weighs about five ounces and is approximately five inches long from head to bottom."

What struck me about these websites' comparisons was how quickly the sizes changed from week to week. At week four, your baby is roughly the size of a poppy seed, but by week five, she has ballooned to the size of a sesame seed! This represents an increase in volume of roughly sixteen times in a week.

Perhaps, though, this rapid increase in size shouldn't be so surprising. When the egg is initially fertilized by the sperm, the resulting zygote undergoes sequential rounds of cell division, called cleavage, which allow the number of cells in the developing embryo to increase rapidly. First, it divides into two. Eight hours later these two further subdivide into four, and after eight more hours, four become eight, which soon turn into sixteen, and so on — just like the number of new investors at each level of the pyramid scheme. Subsequent divisions occur almost synchronously every eight hours. Thus, the number of cells grows in proportion to the quantity of cells in the embryo at a given time: the more cells there are, the more new cells are created at the subsequent division. In this case, since each cell creates exactly one daughter cell at each division, the factor by which the number of cells in the embryo increases is two; in other words, the size of the embryo doubles every generation.

During human gestation the period in which the embryo grows exponentially is, thankfully, relatively short. If the embryo were to carry on growing at the same exponential rate for the whole pregnancy, the 840 synchronous cell divisions would result in a superbaby comprising roughly 10^{253} cells. To put that into context, if every atom in the universe were itself a copy of our universe, then the total number of atoms in all these universes would be roughly equivalent to the number of superbaby's cells. Naturally, cell division becomes less rapid as more complex events in the life of the embryo are choreographed. In reality the number of cells an average newborn baby comprises can be approximated at a relatively modest 2 trillion. This number of cells could be achieved in fewer than forty-one synchronous division events.

The Destroyer of Worlds

Exponential growth is vital for the rapid expansion in the number of cells necessary for the creation of a new life. However the astonishing and terrifying power of exponential growth also led nuclear physicist J. Robert Oppenheimer to proclaim, "Now I am become Death, the destroyer of worlds." This growth was not the growth of cells, nor even of individual organisms, but of energy created by the splitting of atomic nuclei.

During World War Two, Oppenheimer was the head of the Los Alamos laboratory, where the Manhattan Project—to develop the atomic bomb—was based. The splitting of the nucleus (tightly bound protons and neutrons) of a heavy atom into smaller constitutive parts had been discovered by German chemists in 1938. Named nuclear fission in analogy to the binary fission, or splitting, of one living cell into two, as occurs to such great effect in the developing embryo. Fission was found either to occur naturally, as radioactive decay of unstable chemical isotopes, or to be induced artificially by bombarding the nucleus of one atom with subatomic particles in a so-called nuclear reaction. In either case, the splitting of the nucleus into two smaller nuclei, or fission products, was concurrent with the release of large amounts of energy in the form of electromagnetic radiation, as well as the energy associated with the movement of the fission products. It was quickly recognized that these moving fission products, created by a first nuclear reaction, could be used to impact further nuclei, splitting more atoms and releasing yet more energy: a so-called nuclear chain reaction. If each nuclear fission produced, on average, more than one product that could be used to split subsequent atoms, then, in theory, each fission could trigger multiple other splitting events. If continued, the number of reaction events would increase exponentially, producing energy on an unprecedented scale. If a material could be found that would permit this unchecked nuclear chain reaction, the exponential increase in energy emitted over the short timescale of the reactions would potentially allow such a *fissile* material to be weaponized.

In April 1939, on the eve of the outbreak of war across Europe, French physicist Frédéric Joliot-Curie (son-in-law of Marie and Pierre and also a Nobel Prize winner in collaboration with his wife) made a crucial discovery. He published in the journal *Nature* evidence that, upon fission caused by a single neutron, atoms of the uranium isotope U-235 emitted on average 3.5 (later revised down to 2.5) high-energy neutrons. This was precisely the material required to drive the exponentially growing chain of nuclear reactions. The "race for the bomb" was on.

With Nobel Prize–winner Werner Heisenberg and other celebrated German physicists working for the Nazis' parallel bomb project, Oppenheimer knew he had his work cut out at Los Alamos. His main challenge was to create the conditions that would facilitate an exponentially growing nuclear chain reaction allowing the almost instantaneous release of the huge amounts of energy required for an atom bomb. To produce this self-sustaining and sufficiently rapid chain reaction, he needed to ensure that enough of the neutrons emitted by a fissioning U-235 atom were reabsorbed by the nuclei of other U-235 atoms, causing them to split in turn. He found that, in naturally occurring uranium, too many of the emitted neutrons are absorbed by U-238 atoms (the other significant isotope, which makes up 99.3 percent of naturally occurring uranium), meaning that any chain reaction dies out exponentially instead of growing. To produce an exponentially growing chain reaction, Oppenheimer needed to refine extremely pure U-235 by removing as much of the U-238 in the ore as possible.

These considerations gave rise to the idea of the *critical mass* of the fissile material. The critical mass of uranium is the minimum amount required to generate a self-sustaining nuclear chain reaction. It depends on a variety of factors. Perhaps most crucial is the purity of the U-235. Even with 20 percent U-235 (compared to the naturally occurring 0.7 percent), the critical mass is still over four hundred kilograms, making high purity essential for a feasible bomb. Even when he had refined sufficiently pure uranium to achieve supercriticality, Oppenheimer was left with the challenge of the delivery of the bomb itself. Clearly he couldn't just package up a critical mass of uranium in a bomb and hope it didn't explode. A single, naturally occurring decay in the material would trigger the chain reaction and initiate the exponential explosion.

With the specter of the Nazi bomb-developers constantly at their backs, Oppenheimer and his team came up with a hastily developed idea for the delivery of the atomic bomb. In their "gun-type" method, one subcritical mass of uranium was fired into another, using conventional explosives, to create a single supercritical mass. The chain reaction would then be kicked off by a spontaneous fission event emitting the initiating neutrons. The separation of the two subcritical masses ensured that the bomb would not detonate until required. The high levels of uranium enrichment achieved (around 80 percent) meant that only twenty to twenty-five kilograms were required for criticality. But Oppenheimer couldn't risk the failure of his project ceding the advantage to his German rivals, so he insisted on much larger quantities.

By the time enough pure uranium was finally ready, the war in Europe was already over. However, the war in the Pacific region raged on, with Japan showing little sign of surrender despite significant military disadvantages. Understanding that a land invasion of Japan would significantly increase the Americans' already heavy casualties, General Leslie Groves, director of the Manhattan Project, issued the directive authorizing the use of the atomic bomb on Japan as soon as weather conditions permitted.

After several days of poor weather, caused by the tail end of a typhoon, on August 6, 1945, the sun rose in blue skies above Hiroshima. At 7:09 in the morning an American plane was spotted in the skies above Hiroshima and the air-raid warning sounded loud across the city. Seventeen-year-old Akiko Takakura had recently taken a job as a bank clerk. On her way to work as the siren sounded, she took refuge with other commuters in the public air-raid shelters strategically positioned around the city.

Air-raid warnings were not uncommon in Hiroshima; the city was a strategic military base, housing the headquarters of Japan's Second General Army. So far, though, Hiroshima had largely been spared the firebombing that rained down on so many other Japanese cities. Little did Akiko and her fellow commuters know, but Hiroshima was being artificially preserved so that the Americans might measure the full scale of the destruction caused by their new weapon.

At half past seven the all-clear was sounded. The B-29 flying overhead was nothing more sinister than a weather plane. As Akiko emerged from her air-raid shelter, along with many of the others, she breathed a sigh of relief: there would be no air raid this morning.

Unbeknownst to Akiko and Hiroshima's other citizens, as they continued on their journeys to work, the B-29 radioed in reports of clear skies above Hiroshima to the *Enola Gay*—the plane carrying the gun-type fission bomb known as the Little Boy. As children made their way to school and workers continued on their everyday routines, heading for offices and factories, Akiko arrived at the bank in central Hiroshima where she worked. Female clerks were supposed to arrive thirty minutes before the men to clean their offices before the day began, so by ten past eight Akiko was already inside the largely empty building and hard at work.

At 8:14 a.m., the crosshairs of the T-shaped Aioi Bridge came into the sights of Colonel Paul Tibbets, piloting the *Enola Gay*. The forty-four-hundred-kilogram Little Boy was released and began its six mile descent toward Hiroshima. After free-falling for around forty-five seconds, the bomb was triggered at a height of about two thousand feet above the ground. One subcritical mass of uranium was fired into another, creating a supercritical mass ready to explode. Almost instantaneously the spontaneous fissioning of an atom released neutrons, at least one of which was absorbed by a U-235 atom. This atom in turn fissioned and released more neutrons, which were absorbed in their turn by more atoms. The process rapidly accelerated, leading to an exponentially growing chain reaction and the simultaneous release of huge amounts of energy.

As she wiped the desktops of her male colleagues, Akiko looked out of her window and saw a bright white flash, like a strip of burning magnesium. What she couldn't know was that exponential growth had allowed the bomb to release energy equivalent to 30 million sticks of dynamite in an instant. The bomb's temperature increased to several million degrees, hotter than the surface of the sun. A tenth of a second later, ionizing radiation reached the ground, causing devastating radiological damage to all living creatures exposed to it. A second further on and a fireball, three hundred meters across and with a temperature of thousands of degrees Celsius, ballooned above the city. Eyewitnesses describe the sun rising for a second time over Hiroshima that day. The blast wave, traveling at the speed of sound, leveled buildings across the city, throwing Akiko across the room and knocking her unconscious.

Infrared radiation burned exposed skin for miles in every direction. People on the ground close to the bomb's hypocenter were instantly vaporized or charred to cinders.

Akiko was sheltered from the worst of the bomb's blast by the earthquake-proof bank. When she regained consciousness, she staggered out onto the street. As she emerged, she found that the clear blue morning skies had gone. The second sun over Hiroshima had set almost as quickly as it had risen. The streets were dark and choked with dust and smoke. Bodies lay where they had fallen for as far as the eye could see. Only 260 meters from the hypocenter, Akiko was one of the closest to it to survive the terrible exponential blast.

The bomb itself and the resulting firestorms that spread across the city are estimated to have killed around seventy thousand people, fifty thousand of whom were civilians. The majority of the city's buildings were also completely destroyed. Oppenheimer's prophetic musings had come true. The justification for the bombings of both Hiroshima and, three days later, Nagasaki, as necessary measures to end the war, is still debated to this day.

The Nuclear Option

Whatever the rights and wrongs of the atomic bomb, the greater understanding of the exponential chain reactions caused by nuclear fission that was developed as part of the Manhattan Project gave us the technology required to generate clean, safe, low-carbon energy through nuclear power. One kilogram of U-235 can release roughly 3 million times more energy than burning the same amount of coal. Despite evidence to the contrary, nuclear energy suffers from a poor reputation for safety and environmental impact. In part, exponential growth is to blame.

On the evening of April 25, 1986, Alexander Akimov checked in for the night shift at the power plant in which he was shift supervisor. An experiment, designed to stress-test the cooling-pump system, was to get underway in a couple hours. As he initiated the experiment, he could have been forgiven for thinking how lucky he was—at a time when the

Soviet Union was collapsing and 20 percent of its citizens were living in poverty—to have a stable job at the Chernobyl nuclear power station.

At around 11:00 p.m., to reduce for the purposes of the test the power output to around 20 percent of normal operating capacity, Akimov remotely inserted a number of control rods between the uranium fuel rods in the reactor core. The control rods absorbed some of the neutrons released by atomic fission, so that these neutrons didn't cause too many other atoms to split. This put a break on the rapid growth of the chain reaction that would be allowed to run exponentially out of control in a nuclear bomb. However, Akimov accidentally inserted too many rods, causing the power output of the plant to drop significantly. He knew that this would cause reactor poisoning—the creation of material, like the control rods, that would further slow the reactor and decrease the temperature, which would lead to more poisoning and further cooling in a self-reinforcing feedback loop. Panicking now, he overrode the safety systems, placing over 90 percent of the control rods under manual supervision and removing them from the core to prevent the debilitating total shutdown of the reactor.

As he watched the needles on the indicator gauges rise as the power output slowly increased, Akimov's heart rate gradually returned to normal. Having averted the crisis, he moved to the next stage of the test, shutting down the pumps. Unbeknownst to Akimov, backup systems were not pumping coolant water as fast as they should have been. Although it was initially undetectable, the slow-flowing coolant water had vaporized, impairing its ability both to absorb neutrons and to reduce the heat of the core. Increased heat and power output led to more water flash-boiling into steam, allowing more power to be produced: another, altogether more deadly, positive feedback loop. The few remaining control rods that Akimov did not have under his manual supervision were automatically reinserted to rein in the increased heat generation, but they weren't enough. Upon realizing the power output was increasing too rapidly, Akimov pressed the emergency shutdown button designed to insert all the control rods and power down the core, but it was too late. As the rods plunged into the reactor, they caused a short but significant spike in power output, leading to an overheated core, fracturing some of the fuel rods and blocking further insertion of the control rods. As the heat energy rose exponentially, the power output increased to over ten times the usual operating level. Coolant water rapidly turned to steam, causing two massive pressure explosions, destroying the core and spreading the fissile radioactive material far and wide.

Refusing to believe reports of the core's explosion, Akimov relayed incorrect information about the reactor's state, delaying vital containment efforts. Upon eventually realizing the full extent of the destruction, he worked, unprotected, with his crew to pump water into the shattered reactor. As they worked, crew members received doses of two hundred grays per hour. A typical fatal dose is around ten grays, meaning that these unprotected workers received fatal doses in less than five minutes. Akimov died two weeks after the accident from acute radiation poisoning.

The official Soviet death toll from the Chernobyl disaster was just thirty-one, although some estimates that include individuals who helped in the large-scale cleanup are significantly higher. This is not to mention the deaths caused by the dispersal of radioactive material outside the immediate vicinity of the power plant. A fire that ignited in the shattered reactor core burned for nine days. The fire threw into the atmosphere hundreds of times more radioactive material than had been released during the bombing of Hiroshima, causing widespread environmental consequences for almost all of Europe.

On the weekend of May 2, 1986, for example, unseasonably heavy rainfall lashed the highlands of the UK. Within the falling raindrops were the radioactive products of the fallout from the explosion—strontium-90, cesium-137, and iodine-131. In total, around 1 percent of the radiation released from the Chernobyl reactor fell on the UK. These radioisotopes were absorbed by the soil, incorporated by the growing grass, and then eaten by grazing sheep. The result—radioactive meat.

The Ministry of Agriculture immediately placed restrictions on the sale and movement of sheep in the affected areas, with implications for nearly nine thousand farms and over 4 million sheep. Lake District sheep farmer David Elwood struggled to believe what was happening.

The cloud carrying the invisible, almost undetectable, radioisotopes cast a long shadow over his livelihood. Every time he wanted to sell sheep, he had to isolate them and call in a government inspector to check their radiation levels. Each time the inspectors came they would tell him restrictions would only last another year or so. Elwood lived under this cloud for over twenty-five years, until the restrictions were finally lifted in 2012.

It should, however, have been much easier for the government to inform Elwood and other farmers when radiation levels would be safe enough for them to sell their sheep freely. Radiation levels are remarkably predictable, thanks to the phenomenon of exponential *decay*.

The Science of Dating

Exponential decay, in direct analogy to exponential growth, describes any quantity that *decreases* with a rate proportional to its current value — remember the reduction in the number of M&M's each day and the waterslide curve that described their decline. Exponential decay describes phenomena as diverse as the elimination of drugs in the body and the rate of decrease of the head on a pint of beer. In particular, it does an excellent job of describing the rate at which the levels of radiation emitted by a radioactive substance decrease over time.

Unstable atoms of radioactive materials will spontaneously emit energy as radiation, even without an external trigger, in a process known as radioactive decay. At the level of an individual atom, the decay process is random — quantum theory implies that it is impossible to predict when a given atom will decay. However, in a material comprising huge numbers of atoms, the decrease in radioactivity is a predictable exponential decay. The number of atoms decreases in proportion to the number remaining. Each atom decays independently of the others. The rate of decay can be characterized by the half-life of a material — the time it takes for half of the unstable atoms to decay. Because the decay is exponential, no matter how much of the radioactive material is present to start with, the time for its radioactivity to decrease by half will always be the same. Pouring

M&M's out on the table each day and eating the M-up sweets leads to a half-life of one day—we expect to eat half of the sweets each time we pour them out of the bag.

The phenomenon of exponential decay of radioactive atoms is the basis of radiometric dating, the method used to date materials by their levels of radioactivity. By comparing the abundance of radioactive atoms to that of their known decay products, we can theoretically establish the age of any material emitting atomic radiation. Radiometric dating has well-known uses, including approximating the age of the Earth and determining the age of ancient artifacts such as the Dead Sea Scrolls. If you ever wondered how on earth they knew that archaeopteryx was 150 million years old or that Ötzi the iceman died fifty-three hundred years ago, the chances are that radiometric dating was involved.

Recently, more accurate measurement techniques have facilitated the use of radiometric dating in "forensic archaeology"—the use of exponential decay of radioisotopes (among other archaeological techniques) to solve crimes. In 2017, radiocarbon dating exposed the world's most expensive whiskey as a fraud. The bottle, labeled as an 1878 Macallan single malt, was proved to be a cheap blend from the 1970s, much to the chagrin of the Swiss hotel that sold a single shot of it for $10,000. In December 2018, in a follow-up investigation, the same lab found that over a third of "vintage" Scotch whiskeys they tested were also fakes. But perhaps the most high-profile use of radiometric dating is in verification of the age of historical artworks.

Before World War Two, only thirty-five paintings by Dutch Old Master Johannes Vermeer were known to exist. In 1937, a remarkable new work was discovered in France. Lauded by art critics as one of Vermeer's greatest works, *The Supper at Emmaus* was quickly procured at great expense for the Museum Boijmans Van Beuningen in Rotterdam. Over the next few years several more, hitherto unknown, Vermeers surfaced. These were quickly appropriated by wealthy Dutchmen, in part in an attempt to prevent the loss of important cultural property to the Nazis.

Nevertheless, one of these Vermeers, *Christ with the Adulteress*, ended up with Hermann Göring, Hitler's designated successor.

After the war, when this lost Vermeer was discovered in an Austrian salt mine, along with much of the Nazis' other looted artwork, a great search was undertaken to find out who was responsible for the sale of the paintings. The Vermeer was eventually traced back to Han van Meegeren, a failed artist whose work was derided by many art critics as derivative of the Old Masters. Unsurprisingly, immediately after his arrest, Van Meegeren was incredibly unpopular with the Dutch public. Not only was he suspected of selling Dutch cultural property to the Nazis—a crime punishable by death—but he also made huge sums of money through the sale and lived lavishly in Amsterdam throughout the war, when many of the city's residents were starving. In a desperate attempt at self-preservation, Van Meegeren claimed that the painting he sold to Göring was not a genuine Vermeer, but one that he himself had forged. He also confessed to the forgeries of the other new "Vermeers," as well as recently discovered works by Frans Hals and Pieter de Hooch.

A special commission set up to investigate the forgeries appeared to verify Van Meegeren's claims, in part based on a novel forgery, *Christ and the Doctors*, which the commission had him paint. By the time Van Meegeren's trial started in 1947, he was hailed a national hero, having tricked the elitist art critics who so derided him and fooled the Nazi high command into buying a worthless fake. He was cleared of collaboration with the Nazis and given a sentence of just a year in prison for forgery and fraud, but died of a heart attack before his sentence began. Despite the verdict, many (especially those who had bought the "Van Meegeren Vermeers") still believed the paintings to be genuine and continued to contest the findings.

In 1967, *The Supper at Emmaus* was reexamined using lead-210 radiometric dating. Despite Van Meegeren's being meticulous in his forgeries, using many of the materials Vermeer would originally have employed, he could not control the methods by which these materials were created. For authenticity he used genuine seventeenth-century canvases and mixed his paints according to original formulas, but the lead he used for his white lead paint had only recently been extracted from its ore. Naturally occurring lead contains radioactive isotope lead-210 and its parent radioactive species (from which lead is created by decay), radium-226. When the lead is extracted from its ore, most of the radium-226 is removed, leaving only small amounts, meaning relatively little new lead-210 is created in the extracted material. By comparing the concentration of lead-210 and radium-226 in samples, one can date the lead paint accurately since the radioactivity of lead-210 decreases exponentially with a known half-life. A far higher proportion of lead-210 was found in *The Supper at Emmaus* than there would have been if it were genuinely painted three hundred years earlier. This established for certain that Van Meegeren's forgeries couldn't have been painted by Vermeer in the seventeenth century, as the lead that Van Meegeren used for his paints had not yet been mined.

Ice Bucket Flu

Had Van Meegeren been around today, it's likely that his work would have been neatly parceled up into a convenient click-bait article entitled something like "Nine Paintings You Won't Believe Aren't the Real Thing," and spread around the internet. Modern-day fakes, such as the doctored photo of multimillionaire presidential candidate Mitt Romney appearing to line up six letter-adorned supporters to read RMONEY instead of ROMNEY, or the photoshopped snap of "Tourist Guy" posing on the viewing deck of the South Tower of the World Trade Center, seemingly unaware of the low-flying plane approaching in the background, achieved the global exposure that viral marketers' dreams are made of.

Viral marketing is the achieving of advertising objectives through a self-replicating process akin to the spread of a viral disease (the mathematics of which we will look into more deeply in chapter 7). One individual in a network infects others, who in turn infect others. As long as each newly infected individual infects at least one other, the viral message will grow exponentially. Viral marketing is a subfield of an area known as memetics, in which a *meme*—a style, behavior, or, crucially, an idea—

spreads between people through a social network, just like a virus. Richard Dawkins coined the word *meme* in his 1976 book, *The Selfish Gene*, to explain the way in which cultural information spreads. He defined memes as units of cultural transmission. In analogy to genes, the units of heritable transmission, he proposed that memes could self-replicate and mutate. The examples he gave of memes included tunes, catchphrases, and, in a wonderfully innocent indication of the times in which he wrote the book, ways of making pots or building arches. Of course, in 1976, Dawkins had not come across the internet in its current form, which has allowed the spread of once unimaginable (and arguably pointless) memes including #thedress, rickrolling, and lolcats.

One of the most successful, and perhaps genuinely organic, examples of a viral marketing campaign was the ALS ice bucket challenge. During the summer of 2014, videoing yourself having a bucket of cold water thrown over your head and then nominating others to do the same, while possibly donating to charity, was the thing to do in the northern hemisphere. Even I caught the bug.

Adhering to the classic format of the ice bucket challenge, after being thoroughly soaked I nominated two other people in my video, whom I later tagged when I uploaded it to social media. As with the neutrons in a nuclear reactor, as long as, on average, at least one person takes up the challenge for every video posted, the meme becomes self-sustaining, leading to an exponentially increasing chain reaction.

In some variants of the meme, those nominated could either undertake the challenge and donate a small amount to the amyotrophic lateral sclerosis (ALS) association or another charity of their choice, or choose to shirk the challenge and donate significantly more in reparation. In addition to increasing the pressure on nominated individuals to participate in the meme, the association with charity had the added bonus of making people feel good about themselves by raising awareness, and promoting a positive image of themselves as altruistic. This self-congratulatory aspect increased the infectiousness of the meme. By the start of September 2014, the ALS association reported receiving over $100 million in additional funding from over 3 million donors. As a result of the funding received during the challenge, researchers discovered a third gene responsible for ALS, demonstrating the viral campaign's far-reaching impact.

In common with some extremely infective viruses such as flu, the ice bucket challenge was also highly seasonal (an important phenomenon, in which the rate of disease spread varies throughout the year, and that we will meet again in chapter 7). As autumn approached and colder weather hit the northern hemisphere, getting doused in ice-cold water suddenly seemed like less fun, even for a good cause. By September, the craze had largely died off. Just like the seasonal flu, though, it returned the next summer and the summer after in similar formats, but to a largely saturated population. In 2015, the challenge raised less than 1 percent of the previous year's total for the ALS association. People exposed to the virus in 2014 had typically built up a strong immunity, even to slightly mutated strains (different substances in the bucket, for example). Tempered by the immunity of apathy, each new outbreak soon died out as each new participant failed, on average, to pass on the virus to at least one other.

Is the Future Exponential?

A parable of exponential growth is told to French children to illustrate the dangers of procrastination. One day, it is noted that an extremely small algal colony has formed on the surface of the local lake. Over the next few days, the colony is found to be doubling its coverage of the surface of the lake each day. It will continue to grow like this until it covers the lake unless something is done. If left unchecked, it will take sixty days to cover the surface of the lake, poisoning its waters. Since the algal coverage is initially so small, with no immediate threat, the algae is left to grow until it covers half the surface of the lake, when it will more easily be removed. The question is then asked, "On which day will the algae cover half of the lake?"

A common answer that many people give to this riddle, without thinking, is thirty days. But, since the colony doubles in size each day, if the lake is half-covered one day, it will be completely covered the next day. The perhaps surprising answer, therefore, is that the algae will cover half the surface of the lake on the fifty-ninth day, leaving only one day to save the lake. At thirty days the algae takes up less than a billionth of the capacity of the lake. If you were an algal cell in the lake, when would you realize you were running out of space? Without understanding exponential growth, if someone told you on the fifty-fifth day, when the algae covered only 3 percent of the surface, that the lake would be completely choked in five days' time, would you believe it? Probably not.

This highlights the way in which we, as humans, have been conditioned to think. Typically, for our forebears, the experiences of one generation were much like those of the last: they did the same jobs, used the same tools, and lived in the same places as their ancestors. They expected their descendants to do the same. However, the growth of technology and social change is now occurring so rapidly that noticeable differences occur within single generations. Some theoreticians believe that the rate of technological advancement is itself increasing exponentially.

Computer scientist Vernor Vinge encapsulated just such ideas in a series of science fiction novels and essays, in which successive technological advancements arrive with increasing frequency until new technology outstrips human comprehension. The explosion in artificial intelligence ultimately leads to a "technological singularity" and the emergence of an omnipotent, all-powerful superintelligence. American futurist Ray Kurzweil attempted to take Vinge's ideas out of the realm of science fiction and apply them to the real world. In 1999, in his book *The Age of Spiritual Machines*, Kurzweil hypothesized "the law of accelerating returns." He suggested that the evolution of a wide range of systems—including our own biological evolution—occurs at an exponential pace. He even went so far as to pin the date of Vinge's "technological singularity"—the point at which we will experience, as Kurzweil describes it, "technological change so rapid and profound it represents a rupture in the fabric of human history"—to around 2045. Among the implications of the singularity, Kurzweil lists "the merger of biological and nonbiological intelligence, immortal software-based humans, and ultra-high levels of intelligence that expand outward in the universe at the speed of light." While these extreme, outlandish predictions should probably have been confined to the realm of science fiction, some technological advances really have sustained exponential growth over long periods.

Moore's Law—the observation that the number of components on computer circuits seems to double every two years—is a well-cited example of exponential growth of technology. Unlike Newton's laws of motion, Moore's Law is not a physical or natural law, so there is no reason to suppose it will continue to hold forever. However, between 1970 and 2016 the law has held remarkably steady. Moore's Law is implicated in the wider acceleration of digital technology, which in turn contributed significantly to economic growth in the years surrounding the turn of the last century.

In 1990, when scientists undertook to map all 3 billion letters of the human genome, critics scoffed at the scale of the project, suggesting that it would take thousands of years to complete at the current rate. But sequencing technology improved at an exponential pace. The complete "Book of Life" was delivered in 2003, ahead of schedule and within its $1 billion budget. Today, sequencing an individual's whole genetic code takes under an hour and costs less than $1,000.

Population Explosion

The story of the algae in the lake highlights that our failure to think exponentially can be responsible for the collapse of ecosystems and populations. One species on the endangered list, despite clear and persistent warning signs, is, of course, our own.

Between 1346 and 1353, the Black Death, one of the most devastating pandemics in human history (we will investigate infectious-disease spread in more detail in chapter 7), swept through Europe, killing 60 percent of its population. The total population of the world was reduced to around 370 million. Since then the global population has increased constantly without abating. By 1800, the human population had almost reached its first billion. The perceived rapid increase in population at that time prompted the English mathematician Thomas Malthus to suggest that the human population grows at a rate that is proportional to its current size. As with the cells in the early embryo or the money left untouched in a bank account, this simple rule suggests exponential growth of the human population on an already-crowded planet.

A trope of many science fiction novels and films (take the recent block-busters *Interstellar* and *Passengers*, for example) is to solve the problems of the world's growing population through space exploration. Typically, a suitable Earth-like planet is discovered and prepared for habitation for the overspilling human race. Far from being a purely fictional fix, in 2017 eminent scientist Stephen Hawking gave credibility to the proposition of extraterrestrial colonization. He warned that humans should start leaving Earth within the next thirty years, to colonize Mars or the Moon, if our species is to survive the threat of extinction presented by overpopulation and associated climate change. Disappointingly, though, if our growth rate continued unchecked, even shipping half of Earth's population to a new Earth-like planet would only buy us another sixty-three years until the human population doubled again and both planets reached satura-tion point. Malthus forecast that exponential growth would render the idea of interplanetary colonization futile when he wrote, "The germs of existence contained in this spot of earth, with ample food, and ample room to expand in, would fill millions of worlds in the course of a few thousand years."

However, as we have already found (remember the bacteria *Strep f.* growing in the milk bottle at the start of this chapter), exponential growth cannot be sustained forever. Typically, as a population increases, the resources of the environment that sustains it become more sparsely dis-tributed, and the net rate of growth (the difference between the birth rate and the death rate) naturally drops. The environment is said to have a "carrying capacity" for a particular species—an inherent maximum sustainable population limit. Darwin recognized that environmental limitations would cause a "struggle for existence" as individuals "compete for their places in the economy of nature." The simplest mathematical model to capture the effects of competition for limited resources, within or between species, is known as the logistic growth model.

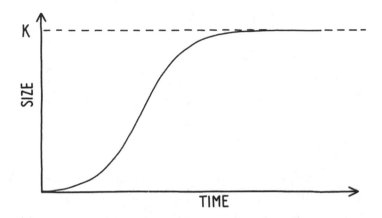

FIGURE 3: The logistic growth curve increases almost exponentially
at first, but then growth slows as resources become a limiting factor and
the population approaches the carrying capacity, K.

In figure 3, logistic growth looks exponential initially as the population grows freely in proportion to its current size, unrestricted by environmental concerns. However, as the population increases, resource scarcity brings the death rate ever closer to the birth rate. The net population growth rate eventually decreases to zero: new births in the population are sufficient to replace those that have died and no more , meaning that the numbers plateau at the carrying capacity. Scottish scientist Anderson McKendrick (one of the earliest mathematical biologists, with whom we will become better acquainted in chapter 7 from his work on modeling the spread of infectious disease) was the first to demonstrate that logistic growth occurred in bacterial populations. The logistic model has since been shown to be an excellent representation of a population introduced into a new environment, capturing the growth of animal populations as diverse as sheep, seals, and cranes.

The carrying capacities of many animal species remain roughly constant, as they depend on the resources available in their environments. For humans, however, a variety of factors, among them the Industrial Revolution, the mechanization of agriculture, and the Green Revolution, have meant that our species has consistently been able to increase its carrying capacity. Although current estimates of the maximum sustainable population of Earth vary, many figures suggest that it is somewhere between 9 billion and 10 billion people. The eminent sociobiologist E. O. Wilson believes that the size of human population that Earth's biosphere can support has inherent, hard limits. The constraints include the availability of fresh water, fossil fuels, and other nonrenewable resources, environmental conditions (including, most notably, climate change), and living space. One of the more commonly considered factors is food availability. Wilson estimates that, even if everyone were to become vegetarian, eating food produced directly rather than feeding it to livestock (since eating animals is an inefficient way to convert plant energy into food energy), the present 1.4 billion hectares of arable land would only produce enough food to support 10 billion people.

If the (near 7.5 billion) human population continues to grow at its current rate of 1.1 percent per year, then we will reach 10 billion inside thirty years. Malthus expressed his fears of overpopulation way back in 1798, when he warned, "The power of population is so superior to the power of the Earth to produce subsistence for man, that premature death must in some shape or other visit the human race." In the timeline of human history, we are now well within the last day left to save the lake.

There are, however, reasons for optimism. Although the human population is still increasing in number, effective birth control and the reduction in infant mortality (leading to lower reproduction rates) have slowed the rate from previous generations. Our growth rate reached a peak of around 2 percent per year in the late 1960s, but is projected to fall below 1 percent per year by 2023. If growth rates had stayed at 1960s rates, it would have taken only thirty-five years for the population size to double. In fact, we only reached 7.3 billion (double the 3.65 billion world population of 1969) in 2016—nearly fifty years later. At a rate of just 1 percent per year we can expect the doubling time to increase to 69.7 years, almost twice as long as the doubling period based on 1969 rates. A small drop in the rate of increase makes a huge difference for exponential growth. By slowing our population growth as we head toward the planet's carrying capacity, we may naturally be beginning to buy ourselves more time. However, exponential behavior may make us, as individuals, feel as if we have less time left than we think.

Time Flies When You're Getting Old

Do you remember, when you were younger, that summer holidays seemed to last an eternity? For my children, who are four and six, the wait between consecutive Christmases seems like an inconceivable stretch of time. In contrast, as I get older, time appears to pass at an alarming rate, with days blending into weeks and then into months, all disappearing into the bottomless sinkhole of the past. When I chat weekly with my septuagenarian parents, they give me the impression that they barely have time to take my call, so busy are they with the other activities in their packed schedules. When I ask them how they fill their week, however, it often seems as if their unrelenting travails might comprise the work of just a single day for me. But then what would I know about competing time pressures? I just have two kids, a full-time job, and a book to write.

I should not be too caustic with my parents, though, because it seems that perceived time really does run more quickly the older we get, fueling our increasing feelings of overburdened time-poverty. In an experiment carried out in 1996, a group of younger people (nineteen to twenty-four) and a group of older people (sixty to eighty) were asked to count out three minutes in their heads. On average, the younger group clocked an almost-perfect three minutes and three seconds of real time, but the older group didn't call a halt until a staggering three minutes and forty seconds, on average. In other related experiments, participants were asked to estimate the length of a fixed period of time during which they had been undertaking a task. Older participants consistently gave shorter estimates for the length of time they had experienced than younger groups. For example, after two minutes of real time, the older group had, on average, clocked less than fifty seconds in their heads, leading them to question where the remaining minute and ten seconds had gone.

This acceleration in our perception of the passage of time has little to do with leaving behind those carefree days of youth and filling our calendars with adult responsibilities. A number of competing ideas explain why, as we age, our perception of time accelerates. One theory notes that our metabolism slows as we get older, matching the slowing of our heartbeats and our breathing. Just as with a stopwatch that is set to run fast, children's versions of these "biological clocks" tick more quickly. In a fixed period of time children experience more beats of these biological pacemakers (breaths or heartbeats, for example), making them feel as if a longer time has elapsed.

A competing theory suggests that our perception of time's passage depends upon the amount of new perceptual information we are sub-jected to from our environment. The more novel stimuli, the longer our brains take to process the information. The corresponding period of time seems, at least in retrospect, to last longer. This argument can explain the movie-like perception of events playing out in slow motion in the moments immediately preceding an accident. In these scenarios, so unfamiliar is the situation for the accident victim that the amount of novel perceptual information is correspondingly huge. It might be that rather than time actually slowing during the event, our recollection of the event is decelerated in hindsight, as our brain records more detailed memories based on the flood of data it receives. Experiments on subjects experiencing the unfamiliar sensation of free fall have demonstrated this.

This theory ties in nicely with the acceleration of perceived time. As we age, we tend to become more familiar with our environments and with life experiences. Our daily commutes, which might initially have appeared long and challenging, full of new sights and opportunities for wrong turns, now flash by as we navigate their familiar routes on autopilot.

It is different for children. Their worlds are often surprising places filled with unfamiliar experiences. Youngsters are constantly reconfiguring their models of the world around them, which takes mental effort and seems to make the sand run more slowly through their hourglasses than for routine-bound adults. The greater our acquaintance with the routines of everyday life, the quicker we perceive time to pass, and generally, as we age, this familiarity increases. This theory suggests that, to make our time last longer, we should fill our lives with new and varied experiences, eschewing the time-sapping routine of the everyday.

Neither of the above ideas explains the almost perfectly regular rate at which our perception of time seems to accelerate. That the length of a fixed period of time appears to reduce continually as we age suggests an "exponential scale" to time. We employ exponential scales instead of traditional linear scales when measuring quantities that vary over a huge range of different values. The most well-known examples are scales for energy waves such as sound (measured in decibels) or seismic activity. On the exponential Richter scale (for earthquakes), an increase from magnitude 10 to magnitude 11 would correspond to a tenfold increase in ground movement, rather than a 10 percent increase as it would do on a linear scale. At one end, the Richter scale captured the low-level tremor felt in Mexico City in June 2018 when Mexican football fans in the city celebrated their goal against Germany at the World Cup. At the other extreme, the scale recorded the 1960 Valdivia earthquake in Chile. The magnitude 9.6 quake released energy equivalent to over a quarter of a million of the atomic bombs dropped on Hiroshima.

If a period of time is judged in proportion to the time we have already been alive, then an exponential model of perceived time makes sense. As a thirty-four-year-old, a year accounts for just under 3 percent of my life. My birthdays seem to come around all too quickly these days. But to ten-year-olds, waiting 10 percent of their life for the next round of presents requires almost saintly patience. To my four-year-old son, the idea of having to wait a quarter of his life until he is the birthday boy again is almost intolerable. Under this exponential model, the proportional increase in age that a four-year-old experiences between birthdays is equivalent to a forty-year-old waiting until he or she turns fifty. When looked at from this relative perspective, it makes sense that time seems only to accelerate as we age.

We commonly categorize our lives into decades—our carefree twenties, our serious thirties, and so on—which suggests that each period should be afforded an equal weighting. However, if time does appear to speed up exponentially, chapters of our life spanning different lengths of time might feel as if they were of the same duration. Under the exponential model, the ages from five to ten, ten to twenty, twenty to forty, and even forty to eighty might all seem equally long (or short). Not to precipitate the frantic scribbling of too many bucket lists, but under this model the forty-year period between forty and eighty, encompassing much of middle and old age, might flash by as quickly as the five years between your fifth and tenth birthdays.

It should be some small compensation, then, for pensioners Fox and Chalmers, jailed for running the Give and Take pyramid scheme, that the routine of prison life, or just the exponentially increasing passage of perceived time, should make their sentences seem to pass very quickly indeed.

In total, nine women were sentenced for their part in the scheme. Although some were forced to pay back part of the money they had made from the scheme, little of the millions of pounds invested in it was recovered. None of this money made its way to the scheme's defrauded investors—the unsuspecting victims who lost everything because they underestimated the power of exponential growth.

From the explosion of a nuclear reactor to the explosion of the human population, and from the spread of a virus to the spread of a viral marketing campaign, exponential growth and decay can play an unseen, but often critical, role in the lives of normal people like you and me. The exploitation of exponential behavior has spawned branches of science that can convict criminals and others that can now, quite literally, destroy worlds. Failing to think exponentially means our decisions, like uncontrolled nuclear chain reactions, can have unexpected and exponentially far-reaching consequences. Among other innovations, the exponential pace of technological advancements has hastened in the era of personalized medicine, in which anyone can have his or her DNA sequenced for a relatively modest sum. This genomics revolution has the potential to lend unprecedented insight into our own health traits, but only, as we will examine in the next chapter, if the mathematics that underpins modern medicine is able to keep pace.

CHAPTER 2

SENSITIVITY, SPECIFICITY, AND SECOND OPINIONS

How Math Makes Medicine Manageable

When I saw the unread email sitting in my in-box, I immediately felt a surge of adrenaline. It started in my stomach and moved down through my arms, causing my fingers to tingle as I clicked open the message. Skimming past the preamble, I clicked immediately on the "View Your Report" link. A browser window opened up; I logged in and clicked on the section headed "Genetic Health Risk." As I scanned the list, I was relieved to see "Parkinson's Disease: Variants not detected," "BRCA1/BRCA2: Variants not detected," "Age-Related Macular Degeneration: Variants not detected." My anxiety subsided as I scrolled past more and more diseases to which I was not genetically predisposed. When I reached the bottom of the list of all clears, my eyes flicked back to one outlying entry I had missed: "Late-Onset Alzheimer's Disease: Increased risk."

When I started writing this book, I thought it would be interesting to investigate the mathematics behind take-at-home genetic tests. So I signed up to 23andMe, probably the best-known personal genomics company out there. How better to understand such results than by taking the test for myself? For a not inconsiderable fee, they sent me a tube in which to collect two milliliters of saliva, which I then sealed and sent back to them; 23andMe promised over ninety reports on my traits, health, and even my ancestry. Over the next few months I didn't give it a second thought, never believing that anything significant would come to light. When the email arrived, however, it suddenly hit me that a comprehensive indication of my future health lay just a couple of clicks away. And then there I was, sitting in front of my computer screen, facing what seemed to be quite serious health implications. As a patient whose meaningful life span seemed to have just been cut short, I naturally panicked. The mathematician inside me, however, kept calm and took over. As I started to rationalize, I wondered, How much could I trust the figures that 23andMe had supplied me with? Does the mathematical methodology currently used to interpret personalized genetic screens stand up to scrutiny?

Whole-genome sequencing, wearable technology, and advances in data science have delivered us into the infancy of the personalized-medicine era. We now have phone apps that can monitor heart rates or estimate aerobic fitness, and at-home tests that claim to be able to diagnose anything from allergies and blood-pressure problems to thyroid problems or even HIV infection. Mathematics is at the heart of this revolution and has played a significant role in transforming all areas of medicine. Math underlies the analysis and interpretation of the ever-expanding range of health tests and screens and provides formulas that are the dispassionate basis for key therapeutic decisions.

In this chapter, we will explore the central role that mathematics now plays in health care. We will investigate the surprising effects of false positive results on the most ubiquitous medical screening programs and get to grips with how tests can be both accurate and yet imprecise at the same time. We will determine whether the formulas that dictate the availability of a particular treatment or that evaluate the impact of our lifestyle choices on our health have a solid grounding in science or whether they are outmoded numerology that need to be discredited and discarded. Ironically, we will draw on centuries-old mathematics to suggest more refined replacements. We will encounter the dilemmas introduced by tools such as pregnancy tests, which give both false positives and false negatives, and see how, in different diagnostic contexts, these incorrect results can be put to good use.

Ordering a DNA screening was my own first taste of the increasingly varied menu offered by the companies serving up personalized medicine. By explaining the methodology currently used to interpret personalized genetic screens, I reinterpret the results of my own DNA tests to understand what my disease risk profile really looks like and determine whether I have bitten off more than I can chew.

What Are the Odds?

In 2007, 23andMe, named after the twenty-three pairs of chromosomes that comprise typical human DNA, became the first company to offer personal DNA testing for establishing a person's ancestry. The following year, thanks to a $4 million investment from Google, 23andMe marketed a saliva test that could estimate how likely you were to suffer from nearly one hundred different conditions, from alcohol intolerance to atrial fibrillation. So comprehensive was their list of traits and so potentially transformative the power of the results that *Time* magazine named the test Invention of the Year.

But the good times didn't last for 23andMe. In 2010, the US Food and Drug Administration (FDA) notified the personal-genomics company that their tests were considered medical devices and therefore required federal approval. In 2013, with 23andMe still lacking this approval, the FDA ordered them to stop providing disease-risk factors until the accuracy of their tests had been verified. 23andMe customers filed a class-action lawsuit, alleging that they had been misled about what the personal-profiling company could deliver. At the height of these troubles, in December 2014, 23andMe launched their health-related services in the UK. Given the controversies, I wondered about the reliability of the tests they might run on my own DNA were I to send them a sample.

Reading about the experiences of thirty-three-year-old web developer Matt Fender in the *New York Times* didn't allay my concerns. As a self-confessed nerd and member of the increasingly large community of the "worried well," Fender is 23andMe's ideal customer. After receiving his profile data and running it through a third-party interpreter, Fender discovered he was positive for the PSEN1 mutation. PSEN1 is an indicator of early-onset Alzheimer's with "complete penetrance," meaning that everyone who has the mutation gets the disease, no ifs, no buts. Unsurprisingly, Fender was alarmed by the thought of losing his ability to think abstractly, to problem solve, and to recall coherent memories. The diagnosis reduced his meaningful life expectancy by at least thirty years.

Unable to take his mind off the implications of the mutation, he sought reassurance. Without any family history of Alzheimer's, Fender struggled to convince geneticists to order a confirmatory follow-up test. Instead, he resorted to a second do-it-yourself genetic test. He sent off another spit kit, this time from Ancestry.com, and awaited the results. They came back five weeks later: negative for PSEN1. Slightly relieved, but even more puzzled than before, Fender convinced a doctor to grant him a clinical assessment, which confirmed Ancestry.com's negative result.

The sequencing technology employed by 23andMe and Ancestry.com, with an error rate of just 0.1 percent, seems incredibly reliable. However, when testing nearly a million genetic variants, it's worth remembering that, even with this low error rate, around a thousand mistakes should be expected. It is worrying, but maybe not surprising, that there can be discord between the results of two independent companies. Perhaps more concerning is the evident lack of post-result support provided. Patients requesting at-home genetic profiles are left to deal with their results in almost complete medical isolation.

After 23andMe gradually gained approval from the FDA for a significantly reduced range of genetic tests, the company relaunched in the United States in 2017, and their home DNA test kit became one of Amazon's top-selling items on Black Friday that year. Despite (or perhaps because of) my misgivings, I ordered a kit and sent off my saliva sample for testing.

When the results came back, the report informed me that 23andMe had detected the epsilon-4 (ε4) variant in one of my two copies of the *apolipoprotein E (APOE)* gene.

In almost every cell of the human body is a nucleus that contains a copy of our DNA—the so-called Book of Life. We inherit these long, twisted ladders of nucleotides in twenty-three pairs of chromosomes, one of each pair coming from each of our parents. Each chromosome in a pair contains copies of the same genes as its partner, whose sequences are similar, but not necessarily exactly the same. For example, 23andMe test for two main variants of the Alzheimer's-associated *APOE* gene, called $\varepsilon 3$ and $\varepsilon 4$. The $\varepsilon 4$ variant is associated with increased risk of late-onset Alzheimer's. Because there are two chromosomes, you can either have one copy of $\varepsilon 4$ (and one copy of $\varepsilon 3$), two copies of $\varepsilon 4$ (and no copies of $\varepsilon 3$), or no copies of $\varepsilon 4$ (and two copies of $\varepsilon 3$)—the number of copies is known as your genotype. Two copies of $\varepsilon 3$ is the most common genotype and is the baseline against which the likelihood of Alzheimer's is judged. The more copies of the $\varepsilon 4$ variant you have, the higher the associated risk of developing Alzheimer's.

But how high is high? Given that 23andMe had found that my genotype comprised one copy of $\varepsilon 4$ and one copy of $\varepsilon 3$, what was my "predicted risk"—the probability of developing the disease? The summary in the report informed me, "On average, a man of European descent with this variant has a 4–7 percent chance of developing late-onset Alzheimer's disease by age 75, and a 20–23 percent chance by age 85." But to be confident of the risks they had predicted for me, I needed to make sure their mathematical analysis was on a sound footing.

The best way to get a handle on the predicted Alzheimer's risk would be to select a huge number of individuals, representative of the population at large, ascertain their genotypes, and then check up on them regularly to see who develops Alzheimer's. With these representative data it would then be easy to compare the risk of getting Alzheimer's given a particular genotype to the risk in the general population—the so-called relative risk. Usually, though, this sort of longitudinal study is prohibitively expensive given the large number of individuals required (especially for rare diseases) and the long time over which they must be observed.

More common, but less powerful, is a case-controlled trial in which a number of individuals already suffering from Alzheimer's are selected, along with a number of "controls"—individuals with similar backgrounds, but without the disease. (We will see in chapter 3 why controlling carefully for the background of individuals is of great importance.) Unlike the longitudinal study, in which participants are selected independently of their disease status, the participants in the case-controlled trial are skewed toward disease carriers, so we aren't able to extract an estimate for the incidence of the disease in the population at large. This means we get a biased prediction of relative risk of the disease. However, these trials do allow us to accurately calculate "odds ratios," which don't require knowledge of the total incidence in the population.

If you've ever been to a greyhound stadium or bet on a horse race, you might remember that the probability of a particular animal winning the race is often expressed as odds. In a given race, an outsider might have odds of 5 to 1 against. This means that if the same race was run a total of six times, we would expect to see this outsider lose five times and win once. The probability of the outsider winning, then, is 1 in 6, or 1/6. The natural way to think about "odds against" is as the ratio of the probability that an event doesn't happen to the probability that it does happen (5/6 to 1/6 in this case, or more simply 5 to 1). Conversely, the favorite for the race might have odds of 2 to 1 on. In sports betting it's traditional to always put the larger number first, so we need to distinguish between odds on and odds against. Odds on, the converse of odds against, expresses the ratio of the probability that an event happens to the probability that it doesn't happen. With odds of 2 to 1 on, if the same race was run three times in total, we might expect the favorite to win twice and lose once. The probability of the favorite winning, then, is 2 in 3 or 2/3, and the probability of it losing is 1/3, which is how we get back to odds on of 2/3 to 1/3, or more simply 2 to 1.

When you hear commentators or bookies describing an "odds-on favorite," it is usually in races with a small number of horses. The phrase, however, is a tautology. Any horse that is odds-on must be the favorite because only one horse in any race is more likely to win than lose. In a race with a larger number of horses it is unusual for a single horse to win more races than it loses. For example, in the UK's most famous horse race, the Grand National, forty horses compete against one another. Even the 2018 winner, Tiger Roll, who started as favorite for (and won) the 2019 race as well, had odds of 4 to 1 against. Because most horses will not be likely to win most of their races, unless explicitly stated otherwise odds at the race course with the largest number first are usually the odds against.

In medical scenarios the opposite is true. Odds are typically expressed as odds on—the probability that an event happens versus the probability that it doesn't—and because we are usually talking about rare diseases (with less than a 50 percent prevalence in the population), the smaller number usually comes first.

To see how to calculate medical odds and the desired odds ratio, let's consider a hypothetical case-controlled study into the effects of having a single ε4 variant (as picked up in my DNA profile) on the incidence of Alzheimer's by age eighty-five, presented in table 1. The odds of developing Alzheimer's by the age of eighty-five, given that you have one copy of the ε4 variant (like me), is the number of people with the disease (100) divided by the number of people without the disease (335): 100 to 335 or, expressed as a fraction, 100/335. By the same logic, drawing figures from the second row of the table, the odds of developing the disease by age eighty-five if you have two copies of the common ε3 variant are 79 to 956 or 79/956. The odds *ratio*, then, is a comparison of the odds of having the disease given one genotype (one copy of the ε4 variant and one copy of the ε3 variant, for example) versus the odds of having the disease given the most common genotype (two copies of the ε3 variant). For the hypothetical figures given in table 1 the odds ratio is 100/335 divided by 79/956, which works out to be 3.61. Crucially, odds ratios don't require us to know the incidence in the population at large and so can easily be calculated from case-control studies.

	ALZHEIMER'S BY 85	NO ALZHEIMER'S BY 85
ε3/ε4	100	335
ε3/ε3	79	956

TABLE 1: Results from a hypothetical case-controlled study on the impact of a single ε4 variant on the occurrence of Alzheimer's by age eighty-five.

Although the odds ratios themselves don't provide the relative risk (the ratio of the risk of getting the disease with the ε3/ε4 genotype to the risk of getting the disease with the ε3/ε3 genotype), they can be combined with the overall population risk of the disease and the known genotype frequencies to find the disease probability for a given genotype. This calculation is not trivial. Indeed, there is not even a unique way to do the calculation. I tried to replicate the late-onset Alzheimer's risks in my genetic report using the same method as 23andMe and data taken directly from the report or from papers they cited. (In case you're interested in calculating the disease probabilities, I used a nonlinear solver to resolve a system of three coupled equations for three unknown conditional probabilities—the sort of thing I enjoy dirtying my hands with in my day job.) I found small, but potentially significant, discrepancies between my figures and theirs. My calculations suggested that I should view the precision of 23andMe's figures with some skepticism.

My conclusion was reinforced when I came across the findings of a 2014 study that investigated the risk-calculation methods of three of the leading personal-genomic companies, including 23andMe. The authors found that differences in the overall population risk, the genotype frequencies, and the mathematical formulas used all contributed to significantly different predicted risks between the companies. When predicted risks were used to categorize individuals into elevated, decreased, or unchanged risk categories, the discrepancies became even more stark. The study found that 65 percent of all individuals tested for prostate cancer were placed in contrasting risk categories (elevated or decreased) by at least two of the three companies. In almost two-thirds of cases, one company might have been telling the individual they were healthy while another company told them they had a significantly increased risk of prostate cancer.

Setting aside the potential for error of the genetic tests themselves, I had found an answer to my questions about the reliability of my results: inconsistencies in the mathematical approach mean that numerical risk calculations presented in personal-genomics health reports should be viewed with some skepticism.

A Eureka Moment

Long before the advent of pricey personalized DNA testing and phone apps that measure your mindfulness or keep tabs on your abs came the cheapest, most easily calculable and decidedly low-tech personal diagnostic tool: the body mass index (BMI). An individual's BMI is calculated by measuring their mass in kilograms and then dividing that by the square of their height in meters.

For recording and diagnostic purposes, anyone with a BMI below 18.5 is classified as "underweight." The "normal weight" range extends from 18.5 to 24.5, and the "overweight" classification spans 24.5 to 30. "Obesity" is defined as having a BMI above 30. Although it is difficult to estimate exactly, obesity may be implicated in around 23 percent of US deaths. The trend is mirrored, to a slightly less extreme extent, throughout the world. In Europe, obesity is second only to smoking as a cause of premature death. Obesity in adults and children is on the rise in almost every country, and its prevalence has doubled over the past thirty years. People with an obese BMI are warned of the dangers of potentially life-threatening conditions such as type 2 diabetes, strokes, coronary heart disease, and some types of cancer, as well as the increased risks of psychological problems such as depression. Today, more people in the world die from being overweight than from being underweight.

Given the health implications of a diagnosis of obesity, or even of being overweight, you might have assumed that the metric used to diagnose these conditions, the BMI, would have a strong theoretical and experimental basis. Sadly this is far from the truth. BMI was first cooked up in 1835 by Belgian Adolphe Quetelet, a renowned astronomer, statistician, sociologist, and mathematician but, notably, not a physician. Using some decidedly shaky mathematics, Quetelet concluded, "The weight of developed persons, of different heights, is nearly as the square of the stature." Notably, though, Quetelet derived this statistic from average population-level data and did not suggest that this ratio would hold true for any given individual. Neither did Quetelet suggest that his ratio, which would become known as the Quetelet index, could be used for making inferences on how over or underweight an individual was, less still about the person's health. This development would not come until 1972. In response to unprecedented levels of obesity, American physiologist Ancel Keys (who would later make the link between saturated fat and cardiovascular disease) undertook a study to find the best indicator of excess weight. He came up with the same ratio of mass to height squared as Quetelet and argued that the measure would be a good indicator of obesity in the population.

Theoretically, individuals who are overweight have a higher mass than their height would suggest and hence a higher BMI. Underweight people would have a correspondingly lower BMI. Keys's BMI formula gained popularity because it was so simple. As we became more overweight as a species and detrimental health outcomes began to be definitively associated with obesity, epidemiologists began to use BMI as a way to track risk factors associated with being overweight. In the 1980s, the World Health Organization, the UK's National Health Service (NHS), and the United States' National Institutes of Health (NIH) all officially adopted the single-value BMI to define obesity for all individuals. Insurance companies on both sides of the Atlantic now routinely use BMI to determine premiums and even whether they will insure an individual at all.

While it's true that fatter people typically have a higher BMI, perhaps unsurprisingly this phenomenological catchall does not work for everybody. The main problem with BMI is that it can't distinguish between muscle and fat. This is important because excess body fat is a good predictor of cardiometabolic risk. BMI is not. If the definition of obesity were instead based on high-percentage body fat, between 15 and 35 percent of men with non-obese BMIs would be reclassified as obese. For example, "skinny-fat" individuals, with low muscle but high levels of body fat and consequently normal BMI, fall into the undetected "normal-weight obesity" category. A recent cross-population study of forty thousand individuals found that 30 percent of people with BMI in the normal range were cardiometabolically unhealthy. The obesity crisis may be much worse than our BMI-based figures suggest. However, BMI both under-

and over-diagnoses obesity. The same study found that up to half of the individuals that BMI classified as overweight and over a quarter of BMI-obese individuals were metabolically healthy.

These incorrect classifications have implications for the way in which we measure and record obesity at a population level. Perhaps more worryingly, though, diagnosing healthy individuals as overweight or obese based on their BMI can also have detrimental effects on their mental health. As a teenager, journalist and author Rebecca Reid battled eating disorders. She cites a biology lesson in which she was taught how to measure BMI as a major trigger point for her struggles. Despite previously being content with her body, when Rebecca measured her BMI, she was labeled overweight. She became so obsessed with the metric that she began a strict diet and exercise program, which saw her lose ten pounds in just a few weeks. Once, she passed out alone in her bedroom while trying to restrict herself to just four hundred calories a day. When not dieting, she punished herself by overeating and then making herself throw up to compensate. Rather than its being a gentle reminder to encourage her to exercise more, Rebecca describes being placed in the overweight category as a "confidence shattering claxon." Ironically, irrespective of their body shape and size, individuals recovering from eating disorders are routinely classified as "recovered" when their BMI reaches 19—just inside the "healthy" range. After taking the incredibly difficult step of admitting to themselves they have a problem and seeking help, some eating-disorder sufferers have even been denied support based on their "healthy" BMIs.

BMI is clearly not an accurate indicator of health at either end of the scale. Instead, it would be more useful to directly measure the percentage of body fat, which is so closely linked to cardiometabolic health outcomes. To do that we need to borrow a two-thousand-year-old idea from the ancient city-state of Syracuse on the island of Sicily.

Around 250 BCE Archimedes, the preeminent mathematician of antiquity (and conveniently a local), was asked by Hiero II, king of Syracuse, to resolve a contentious issue. The king had commissioned a metalsmith to make him a crown of pure gold. After receiving the finished crown and hearing rumors that the metalsmith was less than honest, the king worried that he had been cheated and that the metalsmith had used an alloy of gold and some other cheaper, lighter metal to cut costs. Archimedes was charged with figuring out if the crown was a dud without taking a sample from it or otherwise disfiguring it.

The illustrious mathematician realized he would need to calculate the density of the crown. If the crown was less dense than pure gold, then he would know that the metalsmith had cheated. The density of pure gold was easily calculated by taking a regularly shaped gold block, working out the volume, and then weighing it to find its mass. Dividing the mass by the volume gave the density. So far so good. If Archimedes could just repeat the procedure with the crown, he'd be able to compare the two densities. Weighing the crown was easy enough, but difficulties arose when trying to work out its volume, because of its irregular shape. This problem stumped Archimedes for some time, until one day he took a bath. As he got into his extremely full tub, some of the water overflowed. As he wallowed, he realized that the volume of water that overflowed from a completely full bath would be equal to the submersed volume of his irregularly shaped body. Immediately he had a method for determining the volume, and hence the density, of the crown. Vitruvius tells us that Archimedes was so happy with his discovery that he jumped straight out of the bath and ran naked and dripping down the street shouting, "Eureka!" (I have found it!)—the original eureka moment.

Even today, Archimedes's "displacement" method is used to calculate the volume of irregularly shaped objects. If you were thinking of starting a health drive, you might use it to work out how much smoothie a combination of irregularly shaped fruit and vegetables would make when blended. Alternatively, by blowing as much air as you can into an empty airtight bag and then sealing and submersing it in water, you can use Archimedes's principle to estimate your lung capacity a few weeks into your new exercise program.

Sadly, despite the utility of the displacement method described in the common retelling of the story, this is probably not how Archimedes solved the problem. His measurements of the volume of water displaced by the crown would have had to have been unfeasibly accurate. Instead, Archimedes likely used a related idea from hydrostatics, which would later become known as Archimedes's principle.

The principle states that an object placed in a fluid (a liquid or a gas) experiences a buoyant force equal to the weight of fluid it displaces. That is, the larger the submersed object, the more fluid it displaces, and consequently the larger the upward force it experiences to counteract its weight. This explains why extremely large cargo ships float, providing the weight of the ship and its cargo is less than the weight of water they displace. The principle is also closely related to the property of density—the mass of an object divided by its volume. An object whose density is greater than that of water weighs more than the water it displaces, so the buoyant force is not enough to counteract the object's weight. Consequently, the object sinks.

Using this idea, all Archimedes needed to do was to take a pan balance with the crown on one side and an equal mass of pure gold on the other. In air, the pans would balance. However, when the scales were placed underwater, a fake crown (which would be larger in volume than the same mass of denser gold) would experience a larger buoyant force as it displaced more water, and its pan would consequently rise.

This principle from Archimedes is used to accurately calculate body fat percentage. A subject is first weighed in normal conditions, then reweighed while sitting completely submerged on an underwater chair attached to a set of scales. The differences in the dry and underwater weight measurements can be used to calculate the buoyant force acting on the individual while underwater, which can in turn be used to determine the person's volume, given the known density of water. This volume, in conjunction with figures for the density of fat and lean components of the human body, can be used to estimate the body fat percentage and provide more accurate assessments of health risks.

False Alarms

BMI is just one of a huge number of different mathematical tools that are used routinely throughout the practice of modern medicine. Others range from simple fractions for calculating drug doses to complex algorithms for reconstructing images from CAT scans. Math is being used routinely on the front lines in hospitals to save lives. As we will see shortly, one particularly important place where math is starting to have an impact is in the reduction of false alarms in the intensive care unit (ICU).

False alarms typically refer to an alarm triggered by something other than the expected stimulus. A staggering 98 percent of all burglar-alarm activations in the United States are thought to be false alarms. This prompts the question, Why have an alarm at all? As we get used to incorrect alerts, we become more reluctant to investigate their causes.

Burglar alarms are by no means the only warnings with which we have become overfamiliar. When the smoke detector goes off, we are usually already opening a window and scraping the soot off our toast. When we hear a car alarm outside, very few of us will even get off the sofa and stick our heads outside to investigate. When alarms become an inconvenience rather than an aid, and when we no longer trust their output, we are said to be suffering alarm fatigue. This is a problem because situations in which alarms become so routine that we ignore them, or disable them completely, can be less sensible than not having the alarm in the first place, as the Williams family found out to their great cost.

Michaela Williams spent much of her junior year at high school dreaming of becoming a fashion designer. She had been suffering with long-lasting, frequent, and painful sore throats. Although adolescents are more prone to complications from tonsillectomies than children are, Michaela and her family decided she should undergo the surgery to improve her quality of life. Three days after her seventeenth birthday, she checked in as an outpatient to her local surgical center. After a routine procedure, which took less than an hour, Michaela was taken to the recovery room, while her mother was told that the operation had been a success and that

Michaela could go home later that day. To ease her discomfort in the recovery room, Michaela was given fentanyl, a powerful opioid painkiller. Among the known, but relatively infrequent, side effects of fentanyl is respiratory depression. To be on the safe side, the nurse hooked Michaela up to a monitor measuring her vital signs before going to check on other patients. Despite having the curtains drawn around her, the monitor would rapidly alert the nurse to any deterioration in Michaela's condition.

Or it would have, had the monitor not been muted.

While looking after several patients simultaneously in the recovery room, persistent false alarms had been a nuisance preventing the nurses from doing their jobs efficiently. Having to stop a procedure on one patient to reset an alarm for another was not only costing nurses vital time but also disrupting their concentration. So the nurses had devised a simple solution to allow them to continue their tasks uninterrupted. Routine practice in the recovery room was to turn down, or even completely mute, the monitors to avoid the persistent false alarms.

Shortly after the curtains were drawn around Michaela, the fentanyl caused her breathing to drop drastically. The alarm highlighting hypoventilation was triggered, but no one could see the flashing light through the curtain, and certainly no one could hear it. As Michaela's oxygen levels continued to fall, her neurons began to fire uncontrollably, setting off a chaotic electrical storm that caused irreparable damage to her brain. By the time she was next checked, twenty-five minutes after she was given the fentanyl, she was so severely brain damaged that all chances of survival were gone. She died fifteen days later.

For patients such as Michaela, who are recovering from operations, or who have to spend time in intensive care, having their vital signs monitored with automated alarms that detect everything from heart rate and blood pressure to blood oxygenation and intercranial pressure has obvious benefits. Typically, when the monitor registers above or below a given threshold, the alarm is triggered. However, approximately 85 percent of automated warnings in ICUs are false alarms.

Two factors cause these high rates of false alarms. First, for obvious reasons, alarms in ICUs are set to be extremely sensitive: thresholds for alarms are set deliberately close to normal physiological levels to ensure that even the slightest abnormalities are highlighted. Second, rather than requiring a sustained abnormal signal, alarms are triggered the instant a signal crosses a threshold. When combined, the slightest rise in blood pressure, for example, even for an instant, is enough to trigger the alarm. While this spike could be indicative of dangerous hypertension, it is far more likely to be caused by natural variation or noise in the measuring equipment. However, if blood pressure stayed high for a sustained period, we would be less likely to attribute this to a measuring error. Fortunately, mathematics has a simple way to solve this problem.

In the solution, known as filtering, a signal at a given point is replaced by the average over its neighboring points. This sounds complicated, but we encounter filtered data all the time. When climate scientists claim, "We have just experienced the warmest year since records began," they are not comparing temperature data day by day. Instead, they might average across all the days of the year, smoothing over fluctuating daily temperatures, giving a result that makes for easier comparison.

Filtering tends to smooth signals out, making spikes less pronounced. When you take a photo with a digital camera in low-light conditions, the long exposures required often result in grainy images. Occasional bright pixels appear in dark areas of the image and vice versa. Since the intensity of pixels in a digital photo is represented numerically, filtering can replace the value of each pixel by the average value of its neighboring pixels, filtering out the noise and giving a smoother resulting image.

We can also use different sorts of averages when filtering. The average we are most familiar with is the *mean*. To find the mean, we add up all the values in a data set and divide by how many values there are. If, for example, we wanted to find the mean height of Snow White and the seven dwarves, we would add together their heights and divide by eight. This average would be skewed by Snow White, whose relative tallness makes her an outlier in the data set. A more representative average would be the *median*. To find the median height of the crew, we line up the dwarves and Snow White in height order (Snow White at the front, Dopey at the back) and take the height of the middle person. Since we have eight (an even number) people in the line, we don't have a single middle person. Instead we take the mean height of the middle two (Grumpy and Sleepy) as our median. By using the median we have successfully removed the outlying height of Snow White, which so biased the mean. For the same reason, the median is often used when presenting data on average income. As evident in figure 4, the high wages of the very well-off individuals in our societies tend to distort the mean — an idea we will encounter again in the misleading mathematics in the courtroom in the next chapter. The median gives us a better idea than the mean of what to expect of a "typical" household's disposable income. It could be argued that Snow White's height or the income of high earners should not be neglected in these statistics, as they are as valid as any other data points in the set. While this may be the case, the point is that neither mean nor median is correct in any objective sense. The different averages are simply useful for different applications.

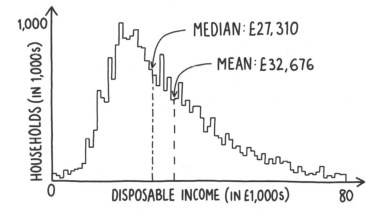

FIGURE 4: The frequency of UK households with a given disposable (after tax) income (in £1,000 blocks) in 2017. The median (£27,310) might be considered a better representation of the "typical" household disposable income than the mean (£32,676).

When filtering a grainy digital image, we want to remove the effects of spurious pixel values. When averaging over neighboring pixel values, mean filtering would modulate, but not completely remove, these extreme values. Conversely, median filtering ignores the values of extremely noisy pixels with impunity.

For the same reason, median filtering is beginning to be used in our ICU monitors to prevent false alarms. Taking the median over a number of sequential readings, alarms are triggered only if thresholds are breached for a sustained (although still short) time, rather than by a one-off spike or dip in a monitor's readout. Median filtering can reduce the occurrence of false alarms in ICU monitors by as much as 60 percent without jeopardizing patient safety.

False alarms are a subcategory of errors known as false positives. A false positive, as the name suggests, is a test result that indicates that a particular condition or attribute is present when it isn't. Typically, false positives occur in binary tests. These are tests with two possible outcomes—positive or negative. With medical tests, false positives result in people who are not sick being told that they are. In the courtroom, false positives are the innocent people convicted of crimes they didn't commit. (We will meet many such victims in the next chapter.)

A binary test can be wrong in two ways. The four possible outcomes of a binary test (two correct and two incorrect) can be read off from table 2 below. As well as false positives, there are also false negatives.

PREDICTED CONDITION	TRUE CONDITION	
	POSITIVE	NEGATIVE
POSITIVE	TRUE POSITIVE	FALSE POSITIVE
NEGATIVE	FALSE NEGATIVE	TRUE NEGATIVE

TABLE 2: The four possible outcomes of a binary test.

With disease diagnostics, you might assume that false negatives are potentially more damaging, since they tell patients that they do not have the disease for which they are being tested when in fact they do have it. We will meet some of the unsuspecting victims of false negatives later in this chapter. False positives can also have surprising and serious implications, but for completely different reasons.

The Big Screen

Take disease screening, for example. Screening is the mass testing, for a particular disease, of asymptomatic people belonging to a high-risk group. For example, in the UK, women over fifty are invited to routine breast screens, as they are at increased risk of developing breast cancer. The occurrence of false positives in medical screening programs is currently a subject of intense debate.

The prevalence of undiagnosed breast cancer among women in the United States may be around 0.2 percent. This means that, at any given time, for every ten thousand women in the United States who have not been diagnosed, we would expect twenty of them to have breast cancer. This doesn't sound high, but this is because, in the majority of cases, breast cancer is detected quickly. In fact, one in eight women will be diagnosed with breast cancer during her lifetime. In the United States, around three in ten of these women are diagnosed late (with regional or distant spread of the tumor). Late diagnosis significantly reduces the chances of long-term survival, supporting the argument that regular mammograms are of vital importance, particularly for women in vulnerable age categories. However, our breast cancer tests have a mathematical problem, of which most people are unaware.

In the spring of 2014, Dominique Berry went to see her doctor to have an irritating skin condition examined. During the consultation at the clinic in downtown Chicago, Dominique's physician noticed a small lump under the skin on her chest. After discovering that Dominique had been ignoring the lump for several weeks, her doctor insisted she have it examined further.

Four days later Dominique arrived at the imaging center for her first-ever mammogram. After a painfully unpleasant few minutes of standing with her breast in a clamp, Dominique retreated to the waiting room, where the nurse eventually found her and told her that the doctor was unhappy with the results. He had discovered a larger mass deeper inside the breast than the lump that had first brought Dominique to the hospital. He requested that she return to undergo a needle biopsy on April 1 — April Fools' Day.

In the intervening two weeks, while Dominique presented an outwardly calm persona at work, on the inside she was falling apart. She agonized about telling her family, but decided she couldn't risk anyone knowing, not even her mother, in case the news leaked out. Instead she relied solely on only her husband for support and threw herself into her work to distract herself from the potential implications of her positive mammogram.

Most patients who undergo mammograms perceive them to be a fairly accurate way of screening for breast cancer. Indeed, for people who have breast cancer, the test will pick this up roughly nine times out of ten. For people who don't have the disease, the results of the test will tell you this correctly nine out of ten times. Knowing these statistics and having received a positive mammography result, even before the biopsy Dominique considered it likely that she had cancer. However, a simple mathematical argument demonstrates that the opposite is true.

The prevalence of undiagnosed breast cancer in women over forty-five — those for whom routine screening is recommended — is slightly higher than in the general female population and can be estimated at around 0.4 percent. The fates of 10,000 such women are broken down in figure 5. We can see that, on average, only 40 of them will have breast cancer, so 9,960 will not. However, one in ten, or 996, of the women who are free of the disease will be given an incorrect positive diagnosis. Compared to the 36 women who are correctly diagnosed as having the disease, this means that a positive test result is correct only in 36 of 1,032 cases or 3.48 percent of the time. The proportion of positive test results that are true positives is known as the precision of the test. Of the 1,032 women to receive a positive result, only 36 of them will actually have breast cancer. To put it another way, if your mammogram comes back positive, the overwhelming likelihood is still that you don't have breast cancer. Despite appearing to be quite an accurate test, the low prevalence of the undiagnosed disease in the population makes it extremely imprecise.

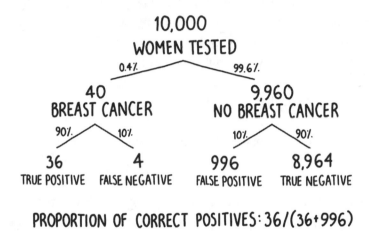

FIGURE 5: Of 10,000 women aged over fifty tested, 36 will be correctly identified as positive whereas 996 will be told they are positive despite not having cancer.

Poor Dominique didn't know this, and neither do many of the women who undergo such tests. Indeed, many doctors are unable to interpret positive mammograms. In 2007, 160 gynecologists were given the following information about the accuracy of mammograms and the prevalence of breast cancer in the population:

- The probability that a woman has breast cancer is 1 percent (prevalence).

- If a woman has breast cancer, the probability that she tests positive is 90 percent.
- If a woman does not have breast cancer, the probability that she nevertheless tests positive is 9 percent.

The physicians were then faced with a multiple-choice question asking them to identify which of the following statements best characterized the chances that a patient with a positive mammogram actually has breast cancer:

A. The probability that she has breast cancer is about 81 percent.
B. Out of ten women with a positive mammogram, about nine have breast cancer.
C. Out of ten women with a positive mammogram, about one has breast cancer.
D. The probability that she has breast cancer is about 1 percent.

The most popular answer among the gynecologists was A—that a positive result in the mammogram will be correct 81 percent of the time (around eight times out of ten). Are they right? Well, we can work out the correct answer by considering the updated decision tree shown in figure 6. With a 1 percent background prevalence, of 10,000 randomly selected women, on average 100 will have breast cancer. Ninety of these will be told correctly that they have the disease by the mammogram. Of the 9,900 women who don't have breast cancer, 891 will be incorrectly told that they do have breast cancer. Of a total of 981 women with a positive result, only 90 of them—or roughly 9 percent—will actually have the disease. Worryingly, the gynecologists massively overestimated the true value. The correct answer was chosen by around one-fifth of respondents, a worse result than if all the doctors had just selected from the four answers at random.

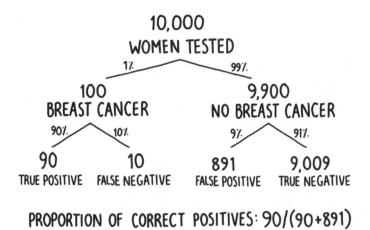

PROPORTION OF CORRECT POSITIVES: 90/(90+891)

FIGURE 6: Of 10,000 hypothetical women in the multiple-choice question, 90 will be correctly identified as positive whereas 891 will be told they are positive despite not having cancer.

As it turned out, when April 1, the day of the biopsy, finally came around, the results showed that the mass in Dominique's breast was an unconcerning benign growth—the whole troublesome affair, in retrospect, turning into the most welcome April Fool she'd ever received. Her travails, though, are typical of the majority of women who receive a positive mammography result. With repeated mammograms, as directed by most screening programs, the chances of receiving a false positive go up. Assuming false positives occur with equal probability of 10 percent (or 0.1) in each test, the correct diagnosis of a true negative occurs with a probability of 90 percent (or 0.9). After seven independent tests the probability of never having received a false positive (0.9 multiplied by itself seven times, or 0.9^7) drops to less than a half (approximately 0.47). In other words, it only takes seven mammograms before an individual free from breast cancer is more likely to have received a false positive than not. With mammograms suggested every year between the ages of forty-five and fifty-four and then every two years after that, women who follow this advice might expect at least one false positive in their lifetime.

The Illusion of Certainty

These high-frequency false positives raise questions about the cost-benefit balance of screening programs. High rates of false positives can have damaging psychological effects and lead patients to delay or cancel future mammograms. However, the problems with screening go beyond simple false positives. Writing in the *British Medical Journal*, Muir Gray, former director of the UK National Screening Programme, admitted, "All screening programmes do harm; some do good as well, and, of these, some do more good than harm at reasonable cost."

In particular, screening can lead to overdiagnosis. Although more cancers are detected through breast screening, many of these are so small or slow growing that they would never be a threat to a woman's health, causing no problems if left undetected. Nevertheless, the C-word induces such mortal fear in most ordinary people that many will, often on medical advice, undergo painful treatment or invasive surgery unnecessarily.

Similar debates surround other mass screening programs, including the smear test for cervical cancer (a disease that we will revisit in chapter 7, when we reconsider the cost-effectiveness and equality of vaccination programs), the PSA test for prostate cancer, and screens for lung cancer. It is important, therefore, that we understand the difference between screens and diagnostic tests. Screening processes can be thought of in analogy to searching for a job. The initial application for a job allows the employer to short-list people for interviews in an efficient way based on a few desirable characteristics. In the same way, screens are designed to cast a wide, less discriminating net across a broad population to identify people who have not yet developed clear symptoms. These are typically less accurate tests, but can be applied in a cost-effective manner to large numbers of people. Employers use more resource-intensive and informative methods, such as assessment centers and interviews, to decide which candidates they will hire. Similarly, once a population of potentially unwell people has been identified through a screen, they can be followed up with more expensive, but more discerning, diagnostic tests to confirm or dismiss the initial screen results. You wouldn't assume you had got the job just because you had been invited to interview. Equally, you shouldn't assume you have a disease because of a positive screen result. When the prevalence of a disease is low, screenings will produce many more false positives than true positives.

The problems caused by false positives in medical screens are, in part, due to our unquestioning attitude toward the accuracy of medical tests. The phenomenon is often known as *the illusion of certainty*. We are so desperate for a definitive answer, one way or another, particularly in medical matters, that we forget to treat our results with the required degree of skepticism.

In 2006, one thousand adults in Germany were asked whether a series of tests gave results that were 100 percent certain. Although 56 percent correctly identified mammograms as having some inaccuracy, the vast majority believed DNA tests, fingerprint analysis, and HIV tests to be 100 percent conclusive, which they demonstrably are not.

In January 2013, journalist Mark Stern spent a week in bed with a fever. He booked an appointment with his new doctor, who decided that the best course of action was to take a blood sample and run a batch of tests. A few weeks later, now feeling better after having taken a course of antibiotics, Mark was alone in his flat in Washington, DC, when the phone rang. His doctor was on the line with Mark's test results. Mark was completely unprepared for the conversation that unfolded.

"Your ELISA test went up to positive," his doctor said, cutting to the chase. "You should go ahead and assume that you have HIV." Despite being unaware that his doctor had even run the ELISA test for HIV (or the follow-up western blot test), when faced with this evidence and the advice of his doctor, Mark had little choice but to reconcile himself to his shock HIV-positive diagnosis. Before ending the call, Mark's doctor suggested Mark come in the next day for confirmatory tests.

That night, Mark and his boyfriend reviewed their previous negative HIV tests from recent months and tried to think of all the subsequent events that could have led to HIV infection. Being in a committed monogamous relationship and practicing safe sex, they found it hard to think of any possibilities. They found it even harder to get to sleep that night.

The next morning, panicked, confused, and exhausted from lack of sleep, Mark reported to the doctor's office. As his doctor drew blood to send away for a confirmatory RNA test, he reiterated his conviction that Mark was HIV positive and suggested a rapid immunoassay test be taken in the office to confirm his belief. As Mark waited out the longest twenty minutes of his life for the test results, he considered what his life with HIV would be like. Although no longer the stigmatized death sentence it once was, he knew a diagnosis would lead him to reevaluate and question many aspects of his life, not least how he had come to be HIV positive in the first place.

At the end of the agonizing wait, no red line had appeared in the results window. Instead, the window allowed a small ray of hope to shine through the clouds onto the troubled landscape of Mark's mind. The test was negative. Two weeks later Mark received the results of the more accurate RNA test—also negative. With a further immunoassay coming back negative, the clouds lifted as his doctor was finally convinced that Mark was HIV negative.

In truth, Mark's original ELISA and western blot tests were ambiguous. His ELISA test did come back with raised levels of antibodies, indicating a positive test. However, at the time he took the test, ELISA had reported false-positive rates of around 0.3 percent. The results of his western blot test—a more accurate test designed to catch such false positives—indicated a lab error. However, Mark's doctor, having never before seen this error, misinterpreted the results. His diagnosis may have been biased by his knowledge that Mark was a gay man, placing him in a high-risk category for HIV. In turn, Mark, blinkered by the illusion of certainty, trusted the judgment of his doctor and the accuracy of the tests.

Two Tests Are Better Than One

The concept of accuracy for two-outcome binary tests is poorly understood by many. From the point of view of those who don't have the disease tested for (which will typically be the vast majority), we could define the "accuracy" of the test as the proportion of these people who are correctly identified as disease-free—the "true negatives." The higher the proportion of true negatives (and thus the lower the rate of false positives), the more accurate the test. The proportion of true negatives is known as the *specificity* of a test. If a test is 100 percent specific, then only people who genuinely have the disease will test positive—there are no false positives.

Even completely specific tests are not guaranteed to identify *everyone* who has the disease. Perhaps we should classify accuracy based on the point of view of the people who actually have the disease. If you were in their shoes, wouldn't you consider it a priority for your disease to be picked up the first time by the test? So perhaps the "accuracy" of a test could be the proportion of "true positives"—the people who have the disease and are identified correctly as such. This proportion is known as the test's *sensitivity*. A test with 100 percent sensitivity would correctly alert all affected patients to their condition.

A test's precision is found by calculating the number of true positives and dividing by the total number of positives, both true and false. The low precision of breast cancer screens, at just 3.48 percent, surprised us earlier in the chapter. The term *accuracy*, however, is typically reserved for the number of true positives and true negatives divided by the total number of people taking the test. This makes sense, as it is the proportion of times the test is giving the correct result, one way or another.

Definitive error rates for the ELISA test for HIV that failed Mark Stern are hard to determine. However, most studies agree on a specificity of around 99.7 percent and a sensitivity closely approximating 100 percent. A negative test result implies the recipient is almost certainly HIV-free, but, on average, three of every thousand people negative for HIV will be given an incorrect HIV-positive diagnosis. The UK has an HIV prevalence of just 0.16 percent. Considering the 1,000,000 randomly selected UK citizens depicted in figure 7, on average 1,600 will be HIV positive and 998,400 will not. Of the 998,400 HIV negative patients undergoing an ELISA test, even with a specificity as high as 99.7 percent, 2,995 will be given an incorrect positive diagnosis. These false positive numbers outweigh the 1,600 true positives by almost two to one. As with breast cancer screening, since the prevalence of HIV is low, and because the

ELISA test lacks a tiny amount of specificity, the proportion of people with positive diagnoses who are genuinely positive (the precision of the test) is low at just over one-third. The accuracy of the test, however, is extremely high. It gives 997,005 correct results (positive or negative) for every million people tested—an accuracy of over 99.7 percent. Even extremely accurate tests can be alarmingly imprecise.

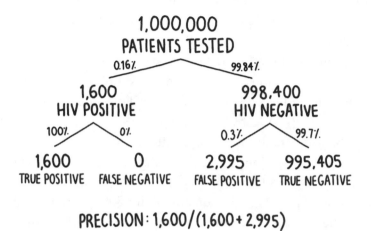

FIGURE 7: Of 1,000,000 people undergoing the ELISA test, 1,600 will be correctly identified as HIV positive, whereas 2,995 will be told they are HIV positive despite not having the disease.

One simple way to improve the precision of a test is simply to run a second test. This is why the first test for many diseases (as we have seen is the case when detecting breast cancer) is a low-specificity screen. It is designed to inexpensively highlight as many potential cases as possible while missing as few as possible. The second test is usually diagnostic and will have a much higher specificity, ruling out the majority of the false positives. Even if a more specific test is not available, a rerun of the same test on all the positive-testing patients can dramatically improve the precision. For the ELISA test, the first attempt effectively increases the prevalence of HIV-positive individuals in the retested population from 0.16 percent to around 34.8 percent: the value of the first test's precision. When we run the test again, as depicted in the decision tree in figure 8, the majority of the original false positives are rooted out by the test's high precision, whereas the true HIV-positive individuals are still correctly identified as such. The precision improves to 1,600/1,609, which equates to roughly 99.4 percent.

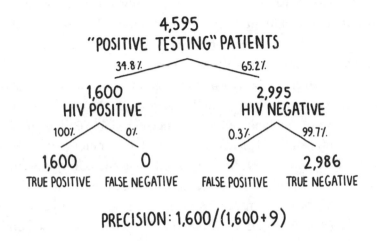

FIGURE 8: Of 4,595 people who originally tested positive, the 1,600 true positives will still be identified as such, but the number of false positives will decrease to just 9.

In theory, a test could be both completely sensitive and completely specific, identifying all the people, and only those people, who have the disease. Such a test could genuinely be claimed to be 100 percent accurate.

Completely accurate tests are not without precedent. In December 2016, a global team of researchers developed a blood test for Creutzfeldt-Jakob disease (CJD). In a controlled trial, the fatal degenerative brain disorder, thought to be caused by eating beef from animals infected with mad cow disease, correctly identified all 32 patients who had the disease (complete sensitivity) with no false positives (complete specificity) among the 391 control patients.

Although there doesn't necessarily have to be a trade-off between sensitivity and specificity, in practice there usually is. False positives and negatives are typically negatively correlated: the fewer false positives the more false negatives and vice versa. In practice, effective tests find a threshold level at which to draw a line between complete specificity and complete sensitivity—a balance somewhere between the two extremes, as close as possible to both.

This trade-off exists because we are typically testing for proxies rather than the phenomena themselves. The test that misdiagnosed Mark Stern as HIV positive does not test for the HIV virus. Rather, it tests for antibodies that the body's immune system raises in an attempt to fight off the virus. However, high HIV-associated antibody loads can be raised by something as innocuous as the flu vaccination. Similarly, most home pregnancy tests do not look for the presence of a viable embryo implanted in the woman's womb. Typically, these tests look for elevated levels of the hormone HCG, produced after implantation of the embryo. Such proxy indicators are often called surrogate markers. Tests can be wrong because markers similar to the surrogate can trigger a positive result.

CJD diagnostic tests, for example, have typically been based on brain scans and biopsies measuring the potential effect on the brain of the faulty proteins that are the root cause of the condition. Unfortunately, the characteristics evaluated by these tests are similar to the characteristics of people with dementia, making clear diagnoses difficult. Rather than looking for subtly different symptoms that could be confused with those of other diseases, the new CJD blood test detects the infectious proteins that always give rise to the disease. This is why the test can be so conclusive: if the malformed proteins are found, then that person has the disease; if not, then the person hasn't. When testing for the root cause of a disease, rather than a proxy, it really is that simple.

Another common reason for the failure of proxy tests is if the surrogate marker is produced by something other than the phenomenon for which we are testing. Anna Howard was just twenty when she woke up feeling sick one morning in June 2016. Although she and her boyfriend of nine months, Colin, were not actively trying for a baby, she decided to take a pregnancy test just in case. She was surprised when the little blue line slowly appeared, as if by magic, as she watched the wand. Neither of them had planned this, but having convinced themselves they would make good parents, Colin and Anna decided to keep the baby and even started to try out some names.

Eight weeks into her pregnancy, Anna started to bleed. Her GP referred her to a hospital for a scan to check that the baby was okay. After the scan, doctors informed Anna that she was miscarrying. They told her to come back the next day for further confirmatory tests. The next day, however, a hormone test, not dissimilar to a home pregnancy test, showed that Anna's levels of HCG, the "pregnancy hormone," were still high enough to indicate a viable pregnancy. Consequently, the doctors informed her that the miscarriage diagnosis was a false alarm.

A week later Anna was bleeding again, and in extreme pain, so she returned to the hospital. This time, fearing an ectopic pregnancy, the doctors inspected Anna's reproductive tract with a fiber-optic camera. They found no evidence of a fetus growing in the wrong place, but the growth in Anna's womb was no fetus either. Rather than a healthy baby, Anna had a gestational trophoblastic neoplasia (GTN)—a cancerous tumor—growing in her uterus. The tumor was growing at roughly the same rate as a fetus and producing HCG, the proxy indicator for pregnancy, deceiving the pregnancy tests, Anna, and the medics alike into thinking her life-threatening cancer was a normal healthy baby.

Although tumors such as Anna's GTN are rare, other types of tumors can also fool pregnancy tests into giving false positives by producing the surrogate indicator HCG. Indeed, the Teenage Cancer Trust states that pregnancy tests have been used to aid the diagnoses of testicular cancer for at least the last decade. Only a small minority of testicular tumors will give a positive result, but since any positive results are known to be false positives for pregnancy, that means that raised HCG levels are extremely likely to be the result of a tumor.

Pregnancy tests are manifestly capable of giving (in some cases quite useful) false positives. However, the levels of HCG in the urine can be so low that these tests are also capable of false negatives. False-negative pregnancy tests, although less common than false positives, can have significant adverse effects on mothers-to-be. In one case, a woman mis-carried after she was sanctioned to undergo a surgical procedure that would never have been undertaken had she known she was pregnant. Another woman's ectopic pregnancy was missed by urine tests, leading to a ruptured fallopian tube and life-threatening blood loss.

In most cases, once pregnancy is well established we abandon the proxy hormonal markers in favor of ultrasound scans, which directly demon-strate the presence of a developing fetus in the womb. However, ultra-sound scans are rarely used to establish pregnancy, but rather to check that the fetus is developing normally. One of the tests that is run at this stage is the nuchal scan, designed to detect cardiovascular abnormalities in the developing fetus. These are typically associated with chromosomal abnormalities such as Patau syndrome, Edwards syndrome, and Down syndrome. For most people, our DNA is bunched into twenty-three numbered pairs of chromosomes. In the three conditions tested for by the nuchal scan, one of the numbered pairs has an extra chromosome, meaning it is actually a chromosome triple or a trisomy.

The nuchal scan is not quite as simple as a binary test. It doesn't pre-dict absolutely whether an unborn child has Down syndrome. Rather, it presents the parents-to-be with an assessment of the risk of the condition. Nevertheless, based on the scan, pregnancies are clearly categorized into high risk and low risk, and this distinction is used when relaying the test results to the parents. If an unborn child is categorized as low risk (less than a 1 in 150 chance) of having Down Syndrome, then no further test-ing is offered, but if the child is in the high-risk category, then the more accurate amniocentesis test is often offered. Fluid containing fetal skin cells is removed, via a needle, from the amniotic sac surrounding the fetus. Piercing the womb and the amniotic sac comes with a risk: between five and ten in every thousand pregnancies that are tested with amniocentesis are subsequently miscarried. However, the increased specificity of the test makes this risk acceptable for many parents-to-be. The test is more accurate than a scan because it detects the extra chromosome in the baby's

DNA (extracted from the fetal skin cells) rather than a proxy marker. It picks out the false positives from the first test and provides the parents of the true positives time to make an informed decision about whether to continue with the pregnancy. The cases that slip through the cracks are the false negatives—the parents who are incorrectly told that their child is at low risk of Down syndrome and are offered no further testing.

Flora Watson and Andy Burrell were one such pair of parents. Back in 2002, having had a scare four weeks into her second pregnancy, Flora decided to pay for the relatively new nuchal test to be carried out privately at ten weeks gestation. After the ultrasound scan, Flora was told that she had an extremely low chance of having a baby with Down syndrome. The likelihood of her having a baby with Down syndrome was compared to that of winning the lottery—about one in 14 million. This is more reassurance than most parents can expect from these sorts of screens. Flora was satisfied that she need not go through with the potentially risky amniocentesis to confirm what the nuchal test had already told her. Instead, she could get on with making excited preparations for the birth of her second child.

Five weeks before her due date, however, Flora noticed that something was wrong. Her unborn baby had started to move less and less. Three weeks later she was in the hospital delivering Christopher. He came quickly, just half an hour after her arrival at the hospital. When he emerged, he was so purple and contorted that Flora thought he was dead. The nurses assured her and Andy that the baby was very much alive, but the news they delivered next would alter the future of their family.

Christopher had Down syndrome. On hearing the news, Andy rushed out of the room and Flora began to cry. What should have been a celebration turned, for them, almost to a wake for the loss of their "healthy baby." For the next twenty-four hours, Flora recalls, "I just couldn't touch him or have him anywhere near me." So Christopher lay alone on the first night of his life, cared for only by the nurses on the ward. When the extended family arrived to meet the new arrival, things got worse. Having brought up another son with learning difficulties, Andy's father urged them to leave Christopher in the hospital. Flora's mother wouldn't even look at the baby.

The life that awaited Flora and Andy when they brought Christopher home was very different from the one they had eagerly anticipated all those months before, when given the results of their nuchal scan. The whole family eventually reconciled themselves to Christopher's condition, but the pressures of looking after a disabled child took their toll. With the time pressures and exhaustion placing too much of a strain on their relationship, Flora and Andy separated. Flora insists she wouldn't have terminated her pregnancy had Christopher's Down syndrome been diagnosed earlier. She still gets angry, though, that she was denied time to adjust and make preparations for her son's condition—a complaint we will hear again in chapter 6 when we discover the perils of automated algorithmic diagnosis. Perhaps the family heartbreak that followed Christopher's birth might have been avoided if it weren't for the false-negative test result.

False positives and false negatives are inescapable. Mathematics and modern technology can help to deal with some of these issues, with tools such as filtering at the forefront of the battle, but other problems we must learn to deal with by ourselves. We should remember that screens are not diagnostic tests and that their results should be taken with a pinch of salt. That's not to say we should completely ignore a positive screen result, but we should wait for the results of a more accurate follow-up before we lose too much sleep. The same is true of personalized genetic tests. The risk categories we are placed in may vary from company to company and can't all be right. As Matt Fender found when faced with a potentially life-limiting Alzheimer's diagnosis, a second test might help to give a more definitive answer.

For some tests, a more accurate version is not available. In these cases, we should remember that even a second run of the same test can dramatically improve the precision of its results. We should never be afraid to ask for a second opinion. Clearly, even doctors—the perceived experts—don't always have the firmest grasp of the figures, despite the illusion of confidence they exude. Before you start to worry yourself unduly based on assertions of a single test, find out its sensitivity and specificity and work out the likelihood of an incorrect result. Question the illusion of certainty and take the power of interpretation back into your own hands. As we'll see in the next chapter, not stopping to question authority figures, especially those exploiting the laws of mathematics, has landed more than one person on the right side of the law, but the wrong side of the prison cell door.

CHAPTER 3

THE LAWS OF MATHEMATICS

———————

Investigating the Role of Mathematics in the Law

———————

Sally Clark walked into the bedroom of her cottage where, minutes before, her husband, Steve, had left their eight-week-old infant son, Harry, asleep. She screamed. Harry lay slumped in his bouncy chair, blue in the face and not breathing. Despite her husband's resuscitation attempts and those of the ambulance crew, Harry was pronounced dead a little over an hour later. A horrible tragedy to befall any new mother. But this was the second time it had happened to Sally Clark.

A little over a year earlier, Steve left their home in the leafy Manchester suburb of Wilmslow to enjoy his department's Christmas dinner. Sally put their eleven-week-old son, Christopher, to bed in his Moses basket on her own that evening. Around two hours later, upon finding Christopher unconscious and gray, she called an ambulance. Despite the crew's efforts, Christopher never woke up. A postmortem carried out three days later attributed his death to a lower respiratory tract infection.

After Harry's death, however, Christopher's postmortem results were reexamined. A cut to the lip and bruising to the legs, which were originally attributed to resuscitation attempts, were given a more sinister interpretation. When Christopher's preserved tissue samples were reanalyzed, evidence of antemortem bleeding in the lungs, missed by the first examination, led the pathologist to suggest smothering.

Harry's postmortem indicated retinal hemorrhaging, spinal damage, and tears in the brain tissue: key indications that Harry might have been shaken to death. Taking both postmortems together, police felt they had enough evidence to arrest Sally and Steve Clark. The crown prosecution service decided not to litigate against Steve (since he had not been present when Christopher died), but Sally was charged with the murders of both of her sons.

The trial that followed would see not one, but four, mathematical mistakes, which would contribute to what is often referred to as Britain's worst miscarriage of justice. By telling Sally's story, in this chapter we will investigate the sometimes tragic but all-too-common courtroom mistakes that can result from mathematical errors. We will meet the participants in similar calamities: the criminal whose conviction was quashed on a mathematical technicality; the judge whose mathematical misunderstanding may have helped to free Amanda Knox, the infamous US student accused of murder. But first, let us hear the case of the French military officer exiled to a brutal prison camp for a crime he didn't commit.

The Dreyfus Affair

Math in the courtroom has a long and not-so-distinguished history. The first notable (mis)use was in a political scandal that would divide the French Republic and that would become known worldwide as the Dreyfus affair. In 1894, a French cleaning lady, working undercover at the German embassy in Paris, recovered a discarded memo. The handwritten message, offering French military secrets to the Germans, led to a witch hunt for a possible German spy in the midst of the French army. The search culminated in the arrest of the French Jewish artillery officer Captain Alfred Dreyfus.

At the resulting court-martial, disliking the opinion of the genuine handwriting expert who suspected Dreyfus was innocent, the French government drafted the unqualified Alphonse Bertillon, the head of the Bureau of Identification in Paris. Confusingly, Bertillon claimed that Dreyfus had written the note to make the writing look like a forgery of his own handwriting—a practice known as autoforgery. Bertillon concocted an abstruse mathematical analysis based on a series of similarities in the pen strokes of repeated polysyllabic words in the memo. He claimed that the probability of a similarity between pen strokes at the starts or ends of any pair of repeated words was 1/5. He calculated that the probability of the four coincidences he had detected among the 26 starts and ends of the 13 repeated polysyllabic words was 1/5 multiplied by itself three times—which works out to be a tiny 16 in 10,000—making their occurrence by chance seem extremely unlikely. Bertillon suggested that the similarities were not coincidences, but "must have been done carefully on purpose and must denote a purposeful intention, probably a secret code." His argument persuaded, or at least baffled, a seven-man jury. Dreyfus was convicted and condemned to life imprisonment in solitary confinement on the lonely penal colony of Devil's Island, several miles off the coast of French Guiana.

Such was the opacity of Bertillon's mathematical argument that neither Dreyfus's defense team nor the government commissioner present at the court understood any of the argument. The presiding judges were likely equally confused, but were too intimidated by the pseudomathematical arguments presented to do anything about it. It took Henri Poincaré, one of the most prodigious mathematicians of the nineteenth century (and whom we will meet again in chapter 6 when we come across his million-dollar problem), to unpick Bertillon's mystifying calculations. Called in over a decade after the original conviction, Poincaré quickly spotted the error in Bertillon's calculations. Instead of calculating the probability of four coincidences in the list of twenty-six starts and ends of the thirteen repeated words, Bertillon had calculated the probability of four coincidences in four words, which naturally is far less likely.

As an analogy, imagine inspecting the person-shaped silhouettes at the end of a target practice at a shooting range. Upon finding ten shots to either the head or chest, you might assume that the gunner was a sharpshooter. When you discover that a hundred or even a thousand shots were fired, you might be less impressed. The same was true for Bertillon's analysis. Four coincidences from four possibilities is indeed unlikely, but there are 14,950 different ways of choosing four options from the twenty-six starts and ends of the words that Bertillon analyzed. The real probability of the four coincidences Bertillon had spotted was roughly eighteen in one hundred, over a hundred times larger than the figure he used to convince the jury. When you take into account that Bertillon would have been equally as happy to find five, six, seven, or more coincidences, we can recalculate the probability of finding four *or more* coincidences as roughly eight in ten. Finding what Bertillon considered an "unusual" number of coincidences is far more likely than not finding them. By exposing Bertillon's miscalculation and arguing that even attempting to apply probability theory to such a question was not legitimate, Poincaré debunked the aberrant handwriting analysis and exonerated Dreyfus. After suffering four years of intolerable conditions on Devil's Island and a further seven years living in disgrace back in France, Dreyfus was finally released in 1906 and promoted to major in the French army. His honor restored, with great magnanimity he went on to serve his country in World War One, distinguishing himself on the front line at Verdun.

Dreyfus's case demonstrates both the power of mathematically backed arguments and the ease with which they can be abused. We will revisit this theme several times in the coming chapters: the tendency, when a mathematical formulation is presented, for heads to nod sagely without asking for further explanation, in deference to the savant who has conjured it into being. The mystery surrounding many mathematical arguments is, in part, what makes them both so impenetrable and, often undeservedly, so impressive. Little is ever challenged. A mathematical form of the illusion of certainty (the phenomenon we met in the previous chapter that leads people to accept unquestioningly the results of medical tests) strikes would-be doubters dumb. The tragedy is that we have failed to learn the lessons from Dreyfus's trial and from numerous other mathematical miscarriages of justice throughout history. As a result, innocent victims have suffered the same fate over and over.

Guilty Until Proven Innocent?

Just as we saw with medical tests in the previous chapter, the law is full of instances in which binary judgments have to be made: right or wrong, true or false, innocent or guilty. The courtrooms of many Western democracies abide by the maxim "innocent until proven guilty"—that the burden of proof should rest with the accuser, not the accused. Almost all countries have done away with the converse presumption, "guilty until proven innocent," a practice bound to result in more false positives and fewer false negatives. However, in some modern-day countries the balance leans toward the presumption of guilt and away from innocence. The Japanese criminal justice system, for example, has a conviction rate of 99.9 percent, with most of these convictions backed up with a confession. For comparison, in 2017–18 the UK's crown court had a conviction rate of 80 percent. Of those charged with a felony in the United States, 68 percent were eventually convicted. Japan's high conviction rate sounds impressive, but is it likely that the Japanese police get the right person in over 999 out of every 1,000 cases?

This high conviction rate is due, in part, to the tough interrogation techniques practiced by Japanese detectives. They are allowed to detain suspects for up to three days without charge, can interrogate suspects without a lawyer present, and are not required to record interviews. These uncompromising techniques are a result of the Japanese legal system, in which establishing motive through confession is a hugely important part of obtaining a guilty verdict. This is compounded by the pressure, applied to the interrogators by their superiors, to extract confessions before physically investigating the evidence related to the case. The interrogators' task is made easier by the seeming willingness of many Japanese suspects to confess to avoid the shame brought upon their families by a high-profile trial. The prevalence of false confessions in the Japanese justice system was highlighted recently by the arrests of four innocent people for malicious internet threats. Before the genuine perpetrator eventually owned up to his crimes, two of the accused had already been coerced into giving false confessions.

Japan's preference for the assumption of guilt is a notable exception. So strong is the "innocent until proven guilty" sentiment in most of the rest of the world that it is enshrined as an international human right in the United Nations' Universal Declaration of Human Rights.

Eighteenth-century English judge and politician William Blackstone even went as far as to quantify the sentiment stating, "It is better that ten guilty persons escape than that one innocent suffer." This view places us firmly in the camp of the false negative, acquitting people who may well have committed a crime, but cannot be proven guilty. Even if there is evidence for the accused's guilt, unless that evidence can convince jurors or judges beyond a reasonable doubt, the accused walks away. In Scottish courts, a third verdict exists that reduces the false-negative rate, if only in name. The "not proven" verdict can be applied to acquittals in which the judge or jury is not sufficiently convinced of the accused's innocence to declare them not guilty. In these cases, although the accused is still acquitted, the verdict itself is not incorrect.

73 Million to One

At Sally Clark's trial in an English courtroom, the conflicting evidence made it difficult for the jury to arrive at a clear-cut guilty or not guilty. Sally was adamant that she had not killed her children. The Home Office pathologist and expert witness for the prosecution, Dr. Alan Williams, contended otherwise. The medical forensic evidence he presented was intricate and confusing for the jury. In the lead-up to the trial, the brain tears, spinal injuries, and retinal hemorrhages that Williams had originally "found" in Harry's postmortem were readily discredited by independent experts. Consequently, the prosecution changed tack and tried to persuade the jury that Harry had been smothered to death, not shaken as was originally claimed. Even Williams changed his mind. Nothing in the medical evidence was clear-cut.

On top of this, the fierce contest between defense and prosecution over the circumstantial evidence surrounding the two deaths fomented further the storm of confusion. The prosecution painted a picture of Sally as a vain and selfish career woman who resented the changes that having children had brought to her lifestyle and her body. A woman so desperate to get back to her prematernal life that she had killed her infant sons.

Why then, argued the defense, did she have a second child so soon after the first? And why had she become pregnant with and then given birth to a third child while the trial was being prepared? The defense argued Sally was clearly distraught about her first son's death. The prosecution twisted the argument, intimating that something in her overt grief was suspicious. The doctor who first saw Christopher when he arrived at the hospital countered that nothing was unusual in Sally's distress after having lost her firstborn child. The arguments went back and forth, adding to the mist clouding the jurors' vision of the truth.

Into this confusion swept expert witness Professor Sir Roy Meadow. While the pathologists argued over the extent of "pulmonary hemorrhage" and "subdural hematomas," Meadow guided the jurors away from the rocks of confusion and toward the safety of a verdict, with a clear beacon of light: a single statistic. He testified that the chance of two children from an affluent family suffering sudden infant death syndrome (SIDS—often referred to as cot death) was one in 73 million. For many of the jurors, this was the most important piece of information they took from the trial: 73 million was too huge a number to ignore.

In 1989, Meadow, at the time an eminent British pediatrician, had edited a book, *ABC of Child Abuse*, in which was the aphorism that came to be known as Meadow's law: "One sudden infant death is a tragedy, two is suspicious and three is murder until proved otherwise." This glib maxim is, however, based on a fundamental misunderstanding of probability. The same misunderstanding with which Meadow would mislead the jury in the case of Sally Clark: the simple difference between dependent and independent events.

The Independence Mistake

Two events are dependent if knowledge of one event influences the probability of the other. Otherwise they are independent. When presented with the probabilities of individual events, common practice is to multiply these probabilities together to find the probability of the combination of the events occurring. For example, the probability that a randomly chosen person from the population is female is 1/2. As illustrated in table 3, of 1,000 people, on average, 500 of them will be female. The probability that a randomly chosen person in the population scores above 110 on a particular IQ test is 1/4. This corresponds to a total of 250 out of the 1,000 people considered in table 3. To find out the probability that someone is female *and* has an IQ over 110, we multiply the probabilities 1/2 and 1/4 together to give a probability of 1/8. This agrees with the 125 (1,000/8) people in the female, high-IQ entry of table 3. Multiplying the two probabilities together to find the joint probability of being female and having a high IQ is perfectly acceptable because IQ and sex are independent: having a particular IQ says nothing about your sex, and being of a particular sex says nothing about your IQ.

IQ	SEX		TOTAL
	MALE	FEMALE	
>110	125	125	250
<110	375	375	750
TOTAL	500	500	1,000

TABLE 3: 1,000 people broken down by IQ and sex.

The prevalence of autism in the United States is roughly 1 per 100, or equivalently, 10 per 1,000. We might assume that to find the probability of being female and autistic we can simply multiply the two probabilities (1/2 and 1/100) together to give a probability of 1/200, or equivalently a prevalence of 5 in every 1,000 people. However, autism and sex are not independent. When we analyze 1,000 randomly chosen people in the population, as in table 4, we see that autism in males (8 in 500) is four times more likely than in females (2 in 500). Only 1 in 5 of those on the autistic spectrum are female. We need this extra piece of information to calculate that the probability that a randomly chosen person from the population is both female and autistic is 2 in 1,000, not 5 in 1,000 as we would have erroneously computed by assuming independence. This illustrates how easy it is to make significant mistakes when we use incorrect assumptions about the independence of events.

AUTISTIC	SEX		TOTAL
	MALE	FEMALE	
YES	8	2	10
NO	492	498	990
TOTAL	500	500	1,000

TABLE 4: 1,000 people broken down by sex and whether or not they have autism.

The events that Meadow was considering in his testimony were the deaths of each of Sally Clark's children by SIDS. For his figures, Meadow used a—then unpublished—report on SIDS for which he had been asked to write the preface. The UK-based report studied 363 SIDS cases from a total of 473,000 live births over three years. As well as providing an overall population rate of SIDS occurrence, the report stratified the data by mother's age, household income, and whether anyone in the household smoked. For an affluent, nonsmoking family such as the Clarks, in which the mother was over the age of twenty-six, there was just one SIDS case for every 8,543 live births.

The first mistake in Meadow's analysis was the assumption that the incidence of SIDS cases were entirely independent events. This led him to calculate the probability of two SIDS deaths in the family by multiplying the figure 8,543 by itself, to arrive at a probability of approximately one occurrence in every 73 million pairs of live births. To justify his assumptions he stated, "There is no evidence that cot deaths run in families, but there is plenty of evidence that child abuse does." With this figure in hand, he suggested that with a birth rate in the UK of around seven hundred thousand a year, such a pair of cot deaths could be expected roughly once every one hundred years.

His assumption was wildly off the mark. Many known risk factors are associated with SIDS, including smoking, premature birth, and bed sharing. In 2001, researchers at the University of Manchester also identified markers in genes related to the regulation of the immune system that put children at increased risk of SIDS. Many more genetic risk factors have since been identified. Children who share the same parents are likely to share many of the same genes and, potentially, the increased risk of SIDS. If one child dies from SIDS, then it is likely that the family has some of the attendant risk factors. Hence the probability of subsequent deaths is greater than the background population average. In reality, it is thought that around one family a year in Britain suffers a second SIDS death.

An analogy for the probability of SIDS deaths is to imagine ten bags of marbles. Nine of these bags each contains ten white marbles. The final bag contains nine white marbles and one black one. This initial state is illustrated on the left of figure 9. On your first go, you choose a bag at random and pick a marble at random from this bag. Since there are one hundred marbles and they are all equally likely to be selected, the probability of choosing the black marble on this first go is one in a hundred. For your second pick you put the first marble you chose back in its bag and draw another from the same bag, completely ignoring the other nine bags. If your first pick was the black marble, then you know you are choosing from the bag containing the black marble for your second pick. This makes the probability of choosing the black marble much higher, at one in ten, rather than one in a hundred. In this scenario, choosing two black marbles (with probability one in a thousand) is much more likely than simply multiplying the original probability of choosing one black marble by itself (to give a probability of one in ten thousand). In the same way, once you have had one child die from SIDS, the probability that the second will also die from SIDS is known to increase.

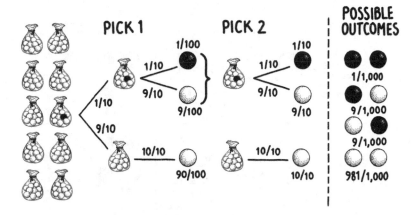

FIGURE 9: A decision tree for finding the probability of picking black or white marbles. To calculate the probability of picking a black or white marble at each attempt, follow the appropriate branches of the tree and multiply the probabilities on each arm. For example, picking a black marble on the first attempt happens with probability 1/100. Once we have chosen a bag on the first attempt, we pick from the same bag on the second attempt. The probabilities of the four two-pick combinations are illustrated to the right of the dashed line.

With SIDS, the risk factors for your family are not randomly chosen when your first child is born, they are preexisting—it could be argued that, right from the start, you are either choosing from the bag with the black marble in it or not. This alternative interpretation is illustrated as a pair of decision trees in figure 10. If you are choosing from the bag with the black marble on both occasions, then the probability of choosing two black marbles increases to one in a hundred. Certainly, simply multiplying the background population risk of SIDS by itself to work out the probability of two SIDS deaths is the wrong thing to do.

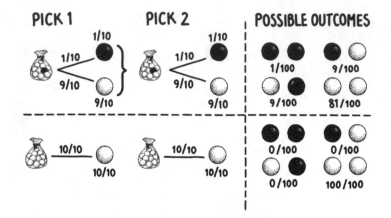

FIGURE 10: Two alternative decision trees in which the bag you choose is predetermined, but still the same bag for both picks. For each tree, the probabilities of the two-pick combinations are illustrated to the right of the dashed line. Clearly, if we are picking from a bag with no black marbles then the only possibility is picking two white marbles.

Meadow's use of the stratified rate of one SIDS case in 8,543 live births had further issues. The report from which he picked this figure also gave a significantly larger overall population risk—1 in 1,303—calculated without stratifying the data by socioeconomic indicators. Meadow chose not to use this alternative figure. Instead, by specifically taking account of the Clarks' background, Meadow produced a number that made a single SIDS case appear far less likely (and because of his mistake of ignoring dependence between the deaths, a double SIDS case less likely still), while neglecting those factors that made it appear more likely. For example, he ignored that both of Sally's children were boys and that SIDS is almost twice as likely in boys as it is in girls. Taking this into account would have undermined the prosecution's argument by making a double SIDS death seem more likely. The prospect that Sally killed both of her children would then appear commensurately less likely.

Although the prosecution's biasing of the statistical evidence by selectively choosing only detrimental background traits may have been considered unethical or misleading, this practice has a deeper problem. The stratification of the data in the original report from which Meadow drew the statistics was implemented to identify high-risk demographics, so as to more efficiently deploy stretched health-care resources. It was never meant to be used to infer the risk of SIDS for a given individual in these groups. The report was a broad-brush investigation into nearly half a million births in the UK, which meant the individual circumstances of each birth could not be investigated in detail. In contrast, the examination of Sally Clark was an extremely detailed investigation into a particular allegation. The prosecution chose only those aspects of Sally and Steve's background that fit with the report and assumed that they could use these to characterize the SIDS risk of the Clarks' children. But this falsely supposes that the characteristics of the individual are the same as those of the population. This is a classic example of what is known as an ecological fallacy.

The Ecological Fallacy

One type of ecological fallacy occurs when we make the unsophisticated assumption that a single statistic can characterize a diverse population. As an example, in the UK in 2010, women had an average life expectancy of eighty-three years. Men, however, had a life expectancy of just seventy-nine years. The overall population life expectancy was eighty-one years. A simple example of an ecological fallacy would be to state that because the average life expectancy of females is higher than that of males, any randomly chosen female will live longer than any randomly chosen male. This fallacy has a special (and appropriate) name — *a sweeping generalization*. Another commonly seen and unsophisticated ecological fallacy based on *increasing* life expectancy is the statement "We are all living longer," which is so often reached for by lazy journalists. Not everyone will live longer than he or she might have expected to previously. Clearly these are naive suggestions at best.

However, ecological fallacies can be more subtle than this. Perhaps it would surprise you to know that despite having a mean life expectancy of just 78.8 years, the majority of British males will live longer than the *overall* population life expectancy of eighty-one years. At first this statement seems contradictory, but it is due to a discrepancy in the statistics we use to summarize the data. The small, but significant, number of people who die young brings down the mean age of death (the typically quoted life expectancy in which everyone's age at death is added together and then divided by the total number of people). Surprisingly, these early deaths take the mean well below the median (the age that falls exactly in the middle: as many people die before this age as after). The median age of death for UK males is eighty-two, meaning that half of them will be at least this age when they die. In this case, the summary statistic presented — the mean age at death of 78.8 years — is a particularly misleading descriptor of the population of British men.

The bell curve, or normal distribution, which can be used to characterize many everyday data sets, from heights to IQ scores, is a beautifully symmetrical curve in which half of the data lies on one side of the mean and half on the other. This implies that the mean and the median — the middle-most data value — tend to coincide for characteristics that follow this distribution. Because we are familiar with the idea that this prominent curve can describe real-life information, many of us assume that the mean is a good marker of the "middle" of a data set. It surprises us when we come across distributions in which the mean is skewed away from the median. The distribution of ages at death for British males, displayed in figure 11, is clearly far from symmetrical. We typically refer to such distributions themselves as skewed.

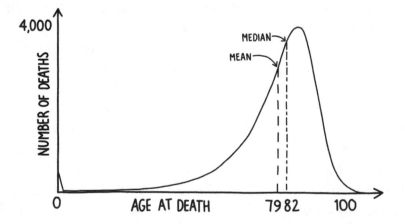

FIGURE 11: The age-dependence of the number of deaths per year for males in Great Britain follows a skewed distribution. The mean age at death is just under 79, while the median age is 82.

As we saw in the previous chapter (where we first introduced the median for the purposes of avoiding false alarms), household income distribution is another statistic for which the median paints a very different picture to the mean. The UK household income distribution shown in figure 4 (page 55), for example, also has a skewed distribution, much like a slightly messier, flipped version of figure 11. The majority of UK households have a low disposable income, but a small, but significant, number of high earners skew the distribution. In the UK in 2014, two-thirds of the population had a weekly income below the "average."

An initially even more surprising example is the old riddle "What is the probability that the next person you meet when walking down the street will have more than the average number of legs?" The answer is "Almost certain." The few people who have no legs or one leg are responsible for a small reduction in the mean, so that everyone with two legs has more than the average. In this case it would be ridiculous to assume that the mean correctly characterized any individual in the population.

Clearly, using the wrong sort of average to describe a population can cause an ecological fallacy. Another type of ecological fallacy, known as Simpson's paradox, occurs when we try to take an average of averages. Simpson's paradox has ramifications in diverse areas from measuring the health of the economy, to understanding voter profiles, and, perhaps most important, in drug development. Imagine, for example, we are in charge of the controlled trial for a new drug, Fantasticol, which is designed to help people reduce their blood pressure. Two thousand people are signed up to the trial, with even numbers of men and women. For control purposes we split them into two groups of one thousand. The patients in group A will receive Fantasticol and those in group B will receive a placebo. At the end of the trial 56 percent (560 out of 1,000) people given the drug are found to have reduced blood pressure, while just 35 percent (350 out of 1,000) in the placebo group have lower blood pressure (see table 5 on page 90). It seems that Fantasticol really does make a difference.

TREATMENT	A: FANTASTICOL	B: PLACEBO
IMPROVEMENT	560	350
NO IMPROVEMENT	440	650
IMPROVEMENT PERCENTAGE	56%	35%

TABLE 5: Fantasticol appears to give a better overall improvement rate than the placebo.

To target the drug properly, we need to know of any sex-specific effects. Consequently, we break the figures down to discover how the drug affects males and females separately. This more detailed breakdown is given in table 6. When we analyze the stratified results, we get a bit of a shock. Among the men in the trial, 25 percent (200 out of 800 in group B) who took the placebo had improved blood pressure readings, but only 20 percent (40 out of 200 in group A) who took Fantasticol improved. Among the women the same trend was apparent; 75 percent (150 out of 200) of women who were taking the placebo improved, in comparison to just 65 percent (520 out of 800) of those taking Fantasticol. For both sexes, a higher proportion of patients taking the placebo improved than of those taking the real drug. Looking at the data this way, it seems that Fantasticol is less effective than the placebo. How can it be that when the data are stratified the trial tells one story, but when amalgamated, they tell the opposite story, and which is correct?

SEX	MALE		FEMALE	
TREATMENT	A: FANTASTICOL	B: PLACEBO	A: FANTASTICOL	B: PLACEBO
IMPROVEMENT	40	200	520	150
NO IMPROVEMENT	160	600	280	50
TOTAL	200	800	800	200
IMPROVEMENT RATE	20%	25%	65%	75%

TABLE 6: When subjects are broken down by sex, patients of both sexes who are taking the placebo do better than patients taking Fantasticol.

The answer lies in something called a "confounding" or a "lurking" variable. In this case the variable is sex. One's sex is important to the results. Throughout the trial, women's blood pressures improved naturally more often than men's did. Because the split of participants' sexes was different in the two groups (800 females and 200 males in drug group A, and 200 females and 800 males in placebo group B) group A benefited significantly from many more women improving naturally, making Fantasticol appear to be more effective than the placebo. Although even numbers of men and women were in the trial, because they were not distributed evenly throughout the two groups, taking the average of the drug's separate success rates for the two sexes (20 percent for men and 65 percent for women) doesn't give the overall 56 percent success rate for Fantasticol that was originally observed in table 5. You can't just average an average.

It's only acceptable to average an average if we are sure we have controlled for the confounding variables. If we had known in advance that sex was one such variable, then we would have known it was necessary to stratify the results by sex to get the true picture of Fantasticol's efficacy. Alternatively, we could have controlled for sex by ensuring we had equal numbers of men and women in each group, as in table 7 (page 92). The improvement rates for men and women taking Fantasticol or the placebo remain the same as in table 6. However, when the results are amalgamated in table 8 (page 92), and we look at the improvement rates for Fantasticol (42.5 percent improvement rate), the drug clearly does worse, not better, than the placebo (50 percent improvement rate). There may, of course, be other confounding variables, such as age or social demographic, for example, that we haven't accounted for.

SEX	MALE		FEMALE	
TREATMENT	A: FANTASTICOL	B: PLACEBO	A: FANTASTICOL	B: PLACEBO
IMPROVEMENT	100	125	325	375
NO IMPROVEMENT	400	375	175	125
TOTAL	500	500	500	500
IMPROVEMENT RATE	20%	25%	65%	75%

TABLE 7: The proportion of men and women who improve under each treatment stays the same as in table 6 when males and females are distributed evenly between the two groups.

TREATMENT	A: FANTASTICOL	B: PLACEBO
IMPROVEMENT	425	500
NO IMPROVEMENT	575	500
IMPROVEMENT RATE	42.5%	50%

TABLE 8: Having controlled for the confounding variable of sex, we see clearly that Fantasticol does worse than the placebo.

Ecological fallacies and well-regulated controls are serious considerations for those who design clinical trials (as we saw in chapter 2 and will see again, but for different reasons, in chapter 4), but they have been known to confound other areas of medicine too. In the 1960s and '70s a curious phenomenon was observed in children whose mothers had smoked while pregnant. Children with low birth weights born to smoking mothers were significantly less likely to die in their first year of life than those born to nonsmoking mothers. Low birth weight had long been associated with higher infant mortality, but it seemed that the mother's smoking during pregnancy was providing some protection to low-birth-weight babies. In reality, it was nothing of the sort. The solution to the paradox lay in a confounding variable.

Although lower birth weight is *correlated* with higher infant mortality, it does not *cause* higher infant mortality. Typically, both can be caused by some other adverse condition: a confounding variable. Both smoking and other adverse health conditions can reduce birth weight and increase infant mortality, but they do so to different extents. Smoking causes many otherwise healthy children to be born underweight. Other causes of low birth weight are typically more detrimental to a child's health, leading to higher infant mortality rates in these cases. The much larger proportion of low-birth-weight children born to smoking mothers, combined with their only slightly increased infant mortality rate, means that a smaller proportion of these children die in their first year than children born with a low birth weight due to some more life-threatening condition.

The ecological fallacy Meadow committed, by pigeonholing the Clarks into the low-risk SIDS category, made their two children's deaths seem far more suspicious than if the higher, population rate of SIDS had been used. Even to use the overall population rate of SIDS would be to commit an ecological fallacy. Arguably, though, the population-level assumption is less partisan, and therefore more appropriate to a situation in which a woman's liberty is on the line. The erroneous assumption of independent SIDS deaths made matters worse.

The Prosecutor's Fallacy

Meadow was not done yet. He was allowed to make an even graver statistical error. This mistake is so common in courtrooms that it is known as the prosecutor's fallacy. The argument starts by showing that, if the suspect is innocent, seeing a particular piece of evidence is extremely unlikely. For Sally Clark this was the assertion that, if she was innocent of killing her two children, the probability of the two infant deaths occurring was as low as one in 73 million. The prosecutor then deduces, incorrectly, that an alternative explanation—the guilt of the suspect—is extremely likely indeed. The argument neglects to take into account any possible alternative explanations, in which the suspect is innocent: the death of Sally's children by natural causes, for example. It also neglects the possibility that the explanation that the prosecution is proposing, in which the suspect is guilty (double infant murder in Sally's case), may be just as unlikely, if not more so than the innocent explanation.

To explain the problems with the prosecutor's fallacy, let's imagine we are investigating a crime. The one piece of evidence we have is the partial license plate number of a car, which must have been that of the perpetrator, seen driving away from the scene. Let's pretend that all license plates are composed of seven numbers, each chosen from the digits 0 to 9. With ten possibilities for each of the seven numbers, that means 10 × 10 × 10 × 10 × 10 × 10 × 10 or 10,000,000 (10 million) such license plates are out on the road. The eyewitness who reported the license plate remembered the first five numbers, but couldn't read the last two. Once these first five digits are specified, we are choosing from a much smaller pool of cars with just two unknown numbers. With ten choices for each of these two unknown digits, that means only a hundred (10 × 10) possible plates are out there with the prescribed first five digits.

A suspect is found whose license plate matches the five digits remembered by the witness. If the suspect is innocent, then only ninety-nine other cars are out there, out of the 10 million cars on the road, whose plates match the first five digits. Therefore, the probability that the witness observed such a license plate if the suspect is innocent is 99/10,000,000, less than one in one hundred thousand (1/100,000). This tiny probability of seeing the evidence if the suspect is innocent seems to overwhelmingly indicate the suspect's guilt. However, to assume so is to commit the prosecutor's fallacy.

The probability of seeing the evidence if the suspect is innocent is not the same as the probability of the suspect being innocent, once that piece of evidence has been observed. Recall that ninety-nine of the hundred cars that match the witness's description do not belong to the suspect. The suspect is just one of a hundred people who drive such a car. The probability of the suspect's guilt given their license plate, therefore, is just one in a hundred—exceedingly unlikely. Other evidence tying the suspect to the area of the crime or eliminating the other cars from being in the area would increase the probability of the suspect's guilt. However, based on the single piece of evidence, the overwhelmingly likely conclusion should be that the suspect is innocent.

The prosecutor's fallacy is only truly effective when the chance of the innocent explanation is extremely small, otherwise it is too easy to see through the fallacious argument. For example, imagine investigating a London burglary. Blood belonging to the perpetrator, discovered at the crime scene, is found to be of the same type as that of a suspect, but no other evidence is retrieved. Only 10 percent of the population share this blood type. So the probability of finding blood of this type at the scene if the accused is innocent (i.e., someone else in the population committed the crime) is 10 percent. The prosecutor's fallacy would be to infer that the probability that the suspect is innocent in light of the blood evidence is also only 10 percent—that the probability of guilt is 90 percent. Clearly, in a city such as London, with a population of 10 million, roughly a million other people (10 percent of the population) will have the blood type found at the crime scene. This makes the probability of the suspect's guilt, based on the blood evidence alone, literally one in a million. Even though finding that blood type was relatively rare (one in ten), because so many people share that blood type, that piece of evidence on its own says little about the guilt or innocence of a suspect whose blood matches.

In the example above the fallacy was relatively obvious. To assume that the probability of innocence could be as low as one in ten based purely on the blood type of an individual in a large population seems absurd. However, in Sally Clark's case, the figures were small enough to make the fallacy quite opaque to a jury untrained in statistics. It is doubtful whether Meadow himself even knew he was committing the fallacy when he stated, "The chance of the children dying naturally in these circumstances is very, very long odds indeed: one in seventy-three million."

The inference that an untrained jury might draw from this statement runs something along these lines: "The deaths of two infants from natural causes is extremely rare; so for a family in which two infants have died, the odds that these deaths were *unnatural* is correspondingly extremely high."

Meadow reinforced this misconception by putting the figure of one in 73 million into a more colorful, but spurious, context. He claimed that the chance of two SIDS deaths in one family was equivalent to betting on the 80-to-1 outsider in the Grand National four years running and winning each time. This made the chance of an innocent explanation for the two infant deaths seem extremely unlikely, with the jury left to assume that the alternative, that Sally had murdered her two children, was therefore extremely likely indeed.

Two children dying from SIDS *is* an extremely unlikely event. This in itself, though, does not provide us with useful information about how likely it is that Sally murdered her children. In fact, the alternative explanation proposed by the prosecution is even more unlikely. Double infant murder has been calculated to be between ten and one hundred times less frequent than double SIDS death. Assuming the latter figure suggests a probability of guilt of just one in a hundred, before any other mitigating evidence has even been considered. However, the probability of double murder was never presented to the jury for comparison. Sally's defense never critically questioned Meadow's statistic, leaving it to stand unchallenged.

After deliberating for two days, on November 9, 1999, the jury found Sally guilty, convicting her by a 10–2 majority. One of the jurors was reported to have confided to a friend that Meadow's statistic was the piece of evidence that swayed the majority of the jury in their verdicts. Sally was sentenced to life imprisonment. As her sentence was read out, Sally looked over to her husband, Steve, who mouthed the words, "I love you." He was her biggest supporter, and he would not stop fighting for her throughout the time she lived in prison, in what she called her "living hell." As she was taken down, she looked back across the gallery and silently spoke his words back to him: "I love you."

The media wasted no time sticking the knife in. The *Daily Mail*'s headline ran "Driven by Drink and Despair, the Solicitor Who Killed Her Babies," while the *Daily Telegraph* proposed, "Baby Killer Was 'Lonely Drunk.'" Sally's reputation on the outside was in tatters, but as a convicted child killer and the daughter of a policeman, life on the inside looked as if it would be hell for her.

Sally spent a year in prison, locked away from her husband and her young son. Her only comforts were the letters she received from strangers who believed she was innocent. On the outside, Steve also maintained his belief in Sally's innocence. After nearly twelve months of hard work, they were finally ready to face the judges again in the court of appeal. The primary basis for the appeal was the inaccuracy of the statistics. Expert statisticians explained to the judges the ecological fallacy in pigeonholing the Clarks into a low SIDS risk category, the erroneous assumption of independence Meadow had made by squaring the probability of a single SIDS death, and the prosecutor's fallacy to which the jury had been subjected.

The presiding judges appeared to understand all of these arguments and took them into consideration. In their summary, they accepted that Meadow's statistics were not accurate but argued that they were only ever supposed to be ballpark figures. The judges believed the prosecutor's fallacy to be so obvious that it should have been objected to by Sally's defense lawyer. That no objections were raised was taken as evidence by the judges that the fallacy was abundantly clear to everyone:

> It is stating the obvious to say that the statement "In families with two infants, the chance that both will suffer true SIDS deaths is 1 in 73 million" is not the same as saying "If in a family there have been two infant deaths, then the chance that they were both unexplained deaths with no suspicious circumstances is 1 in 73 million." You do not need the label "the prosecutor's fallacy" for that to be clear.

The judges concluded that the role of the statistical evidence in the trial was so insignificant that the jury could not possibly have been misled. Far from being the rock offered the jury to cling to in the storm of contradictory medical evidence, the statistics, it seemed, were no more than a drop in the ocean, "a sideshow" as the judges dismissed them. Sally's original conviction was upheld and that same evening she was taken back to prison.

✳   ✳   ✳

Sally Clark's is by no means the only trial in which probability has been misused and misunderstood. In 1990, Andrew Deen was maligned by the same prosecutor's fallacy in his trial for the rape of three women in his native Manchester in the northwest of England. He was convicted and sentenced to sixteen years in prison. At the trial, the prosecuting barrister, Howard Bentham, presented DNA evidence from semen found on one of the victims. Bentham claimed that DNA from a sample of Deen's blood matched the DNA of the semen sample. When he asked the expert witness, "So the likelihood of this being any other man but Andrew Deen is one in three million?," the expert answered, "Yes." The expert added, "My conclusion is that the semen originated from Andrew Deen." Even the judge in his summing up claimed the one in 3 million figure "approximated pretty well to certainty."

In fact, the one in 3 million figure should be interpreted as the probability that a randomly chosen individual from the population at large had a DNA profile matching that of the semen found at the crime scene. Given the roughly 30 million males in the UK at the time, we might expect ten of them to match the profile, dramatically increasing the probability of Deen's innocence from an improbable one in 3 million to a very likely nine in ten. Of course not all of the 30 million males in the UK are possible suspects. However, even if we restrict ourselves to the 7 million people who live within an hour's drive of Manchester's city center, we would still expect at least one other male to match the profile, leaving the odds of Deen's innocence at even: one to one. The prosecutor's fallacy had led the jury to believe Deen was millions of times more likely to be guilty than the evidence actually suggested.

In fact, even the DNA evidence that linked Deen to the crimes was not as convincing as the expert witness had claimed. On appeal, Deen's DNA and the DNA found at the crime scene were shown to be nowhere near as similar as originally thought. Instead of one in 3 million, the probability of a random match with someone other than Deen was actually around one in twenty-five hundred, making Deen's innocence dramatically more likely. Combined with the more than 3 million males in the vicinity of the crime scene, giving over a thousand other potentially matching individuals, the probability of Deen's guilt based on DNA dropped to less than one in a thousand. The revised interpretation of the forensic evidence, and the recognition that both the original judge and the expert witness had committed the prosecutor's fallacy, caused Deen's conviction to be quashed.

Knox and the Knife

Another case in which the understanding of DNA evidence and probability combined to play a pivotal role was that of murdered British student Meredith Kercher. In 2007, Kercher was stabbed to death in the apartment she shared with fellow exchange student Amanda Knox in Perugia, Italy. Two years later, in 2009, Knox and her Italian former boyfriend Raffaele Sollecito were unanimously convicted of Kercher's murder. One piece of evidence presented by the prosecution, which proved crucial in the Knox and Sollecito convictions, was a knife, of a size and shape consistent with some of the wounds inflicted on Kercher. The knife was found in Sollecito's kitchen and had Knox's DNA on the handle, tying both Sollecito and Knox to the weapon. A second sample of DNA was also on the blade of the knife, although it was small, just a few cells. A DNA profile produced from the cells was found to be a positive match for Kercher.

In 2011, Knox and Sollecito appealed against their lengthy prison sentences. Lawyers for the defense primarily focused on discrediting the only evidence that physically tied Knox and Sollecito to the murder—the DNA on the knife.

The genome of almost everyone (the only exceptions being identical siblings)—the readout of all the A's, T's, C's, and G's that characterize the long strings of DNA in each of the person's cells—is unique. If each of the roughly 3 billion base pairs in a person's genome were read off and stored, the resulting sequence would constitute a genuinely unique identifier for that person. A DNA profile used in court or stored in a DNA database, however, is not an exact readout of an individual's whole genome. When DNA profiles were first conceived of, generating a full-genome profile would have required too much data, taken too long, and been too expensive. The comparison of two profiles would also have taken an unfeasible amount of time.

Instead, a DNA profile is produced by analyzing thirteen pairs of regions, known as loci, on different chromosomes of a person's DNA. Each of these regions contains, in part, a "short tandem repeat": a small segment of DNA that repeats numerous times in a row. The loci are polymorphic, meaning the number of repeats can vary significantly between individuals. Indeed, these thirteen loci are specifically chosen because of the astronomically large number of possible repeat patterns that can occur; it is highly improbable that two unrelated people will have the same number of repeats at all thirteen pairs of loci. A DNA profile lists the numbers of repeats at each locus, which collectively can be thought of as an individual's genetic fingerprint. The repeat numbers themselves can be read off from a chart known as an electropherogram. The electropherogram represents the raw DNA sequence and looks a little like the readout of a seismometer (used for measuring earthquakes) with low-level background noise interspersed with peaks at particular positions, corresponding to each of the loci used in the profile. The electropherogram for the sample from the blade of the knife is displayed in figure 12.

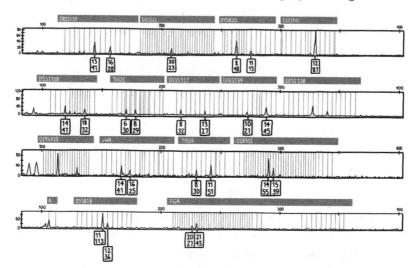

FIGURE 12: The electropherogram from the knife blade DNA sample, allegedly belonging to Kercher. The peaks corresponding to the thirteen loci are labeled. In some instances only one peak is visible, indicating that the same number of repeats for that locus was inherited from both parents. The top number in each box gives the number of repeats. The bottom number gives the strength of the signal. Most signal strengths are lower than the desired minimum of 50.

Creating an individual electropherogram can be likened to recording the results of two rolls of each of thirteen separate dice, in order, each dice with up to eighteen faces. Two randomly chosen individuals' profiles being a perfect match can be thought of as rolling exactly the same sequence twice. Under ideal conditions, the probability of two randomly selected, unrelated individuals' profiles matching is lower than one in a 100 trillion—effectively making the DNA profile a unique identifier. If the peaks on two profiles of the electropherogram match exactly, then it can reasonably be assumed they come from the same person.

Sometimes DNA matches can be ambiguous because the age or quality of the DNA sample means only partial profiles can be recovered—the signal is not attainable at every locus. Partial profiles cannot give such a definitive match. It is also possible, especially for small samples, that the signal that comes through on the electropherogram can be drowned out by the background noise produced during the analysis. For that reason the strength of the signals in the DNA profile must meet accepted standards. The only hope left to Knox's defense was that the DNA sample's signal was too weak.

At the time of the original trial, the chief technical director of the forensic genetic-investigation section of the Rome Police decided that, based on its tiny size, instead of dividing the knife DNA sample into two, she needed to use all of the available DNA to create a sufficiently strong profile. This was not good practice: with two samples a low strength or ambiguous profile can be revalidated using the second sample. As noted in the original trial, the electropherogram possessed clear peaks in all the right places and was an incredibly close match for Kercher's profile. However, as can be read off from the numbered boxes in figure 12, most of the peak heights in the profile fell well short of even the most relaxed of standards. Because the chief technical director had not separated the sample into two, the defense team at the appeal were able to discredit the knife DNA evidence.

In response, the prosecution asked for a small number of cells, missed by the original swab but discovered by independent forensic experts, to be retested to confirm the results of the first test. Presiding judge Claudio

Hellmann rejected the prosecution's requests to have the minuscule sample retested.

On October 3, 2011, the mixed jury of judges and laypeople retired to consider their verdict. They returned, later than expected, to a courtroom whose atmosphere had slowly been building and that was now extremely tense and thick with pent-up emotion. Despite all the evidence that had been reviewed, no one knew which way the pendulum would swing. As the verdicts were read out, Knox collapsed in her chair and began to cry— mixed tears of joy and relief. The jury cleared her of Kercher's murder. In his summarizing "motivations" document, in justifying his refusal to allow the second knife DNA sample to be tested, Judge Hellmann stated, "The sum of two results, both unreliable due to not having been obtained by a correct scientific procedure, cannot give a reliable result." But Leila Schneps and Coralie Colmez, authors of the 2013 book *Math on Trial: How Numbers Get Used and Abused in the Courtroom*, suggest that Judge Hellmann was wrong; sometimes two unreliable tests are better than one.

To understand their argument, imagine, rather than testing DNA for a match, we have a dice to roll. We would like to determine whether the dice is fair, in which case a six should come up one sixth of the time, or if it is weighted, in which case we are told a six should appear 50 percent of the time. Because we don't want to presuppose anything about the situation, assume that, before we carry out our tests, each of these scenarios is equally likely.

We begin by rolling the dice sixty times. If the dice is unbiased, then we would expect to see a six come up ten times on average. If it is weighted, we would expect to see thirty sixes on average. If we find thirty sixes or more in the trial, then we will be confident that the dice is weighted because it would be extremely unlikely for this to happen by chance using an unweighted dice. Similarly, if we find ten or fewer sixes, then we will be confident that the dice is fair. If the number of sixes falls somewhere between ten and thirty, then we will be able to calculate the probability of the dice being weighted by comparing the probability that the number of sixes would come up with the weighted dice, to the probability of seeing that number of sixes with the unweighted dice.

In the experiment, we record the rolls seen in the top half of figure 13 — twenty-one sixes. The probability of seeing this many sixes with an unweighted dice is low, at just 0.000297. With a weighted dice the probability of seeing twenty-one sixes is still quite small, but at 0.00693 it is over twenty times more likely than if the dice is unweighted. The twenty-one sixes are far more likely to have come from a weighted dice than an unweighted one. We can find the combined probability of seeing twenty-one sixes under both of these scenarios by adding their respective probabilities together to give 0.00722. The proportion of this probability that the weighted dice accounts for is 0.00693/0.00722, which gives 0.96. The probability that the dice is weighted, therefore, is 96 percent. Fairly convincing, but perhaps not convincing enough to convict a murderer.

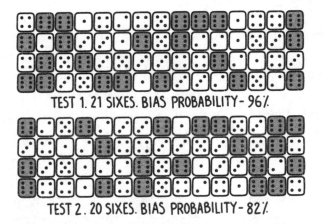

TEST 1. 21 SIXES. BIAS PROBABILITY - 96%

TEST 2. 20 SIXES. BIAS PROBABILITY - 82%

FIGURE 13: Two separate tests of the dice. We roll 21 sixes from 60 rolls on the first test, but only 20 sixes on the second test. The second test seems to undermine the first.

To be sure, we undertake a second test and roll the dice another sixty times. This time, if we count up the sixes in the bottom half of figure 13, we find only twenty. As summarized in table 9 (page 104), the probability of seeing this number of sixes if the dice is unweighted is 0.000780 and, if the dice is weighted, 0.00364 — only around five times more likely now. Although not hugely different to the results of the first test, applying the same calculation gives a slightly less convincing 82 percent chance of the dice being weighted. It seems that undertaking this second test has cast doubt on the results of the first. The second test certainly doesn't seem to confirm our conviction that the dice is weighted beyond a reasonable doubt.

	PROBABILITY GIVEN UNWEIGHTED DICE	PROBABILITY GIVEN WEIGHTED	TOTAL PROBABILITY FOR BOTH SCENARIOS	PROBABILITY DICE IS WEIGHTED
TEST 1	0.000297	0.00693	0.00722	96%
TEST 2	0.000780	0.00364	0.00442	82%
COMBINED	0.00000155	0.000168	0.000170	99%

TABLE 9: The probability of seeing the different numbers of sixes in each of the tests if the dice is fair (column 1) or weighted toward sixes (column 2). The total probability with the two scenarios (column 3) and the probability that the dice is weighted (column 4).

However, when we combine the results, as in figure 14, we find we have rolled the dice 120 times. For an unbiased dice we would expect six to come up 20 times on average. Instead it came up 41 times. The probability of seeing 41 sixes from 120 rolls is just 0.00000155 if the dice is unweighted, whereas, if the dice is weighted, 41 sixes are over one hundred times more likely, at 0.000168. The probability that the dice is weighted given these 41 sixes is, therefore, over 99 percent.

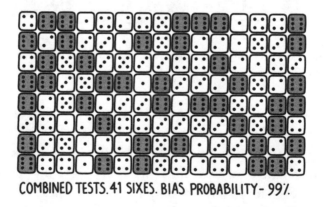

COMBINED TESTS. 41 SIXES. BIAS PROBABILITY – 99%.

FIGURE 14: When the tests are combined we find 41 sixes from a total of 120 rolls. This suggests the overwhelming likelihood that the dice is weighted.

Surprisingly, combining the two less persuasive investigations makes for a much more convincing result than either of the individual tests alone. A similar technique is often employed in the scientific practice of systematic reviews. Systematic reviews in medicine, for example, consider multiple clinical trials, which may not, in themselves, be conclusive about the effectiveness of a given treatment due to the small number of trial participants. When the results of multiple independent trials are combined, however, statistically significant conclusions can often be drawn about the effectiveness of the intervention. Perhaps the best-known use of systematic reviews is in the analysis of alternative medicines (the seeming "positive effects" of which we will explain, in the next chapter, as being primarily caused by mathematical artifacts), for which little funding is available to conduct large-scale clinical trials. By combining multiple seemingly inconclusive tests, systematic reviews have debunked alternative therapies from the use of cranberries to treat urinary tract infections to the use of vitamin C for preventing the common cold.

Similarly, Schneps and Colmez argue that the combination of a pair of potentially inconclusive DNA tests might have provided stronger evidence for the link between Kercher's DNA and the knife in Sollecito's kitchen. Judge Hellmann's decision deprived the court of the opportunity to hear such evidence and consequently denied the world the opportunity to see the effect that evidence might have had on the outcome of the trial.

Blinded by Math

The astronomically small probabilities generated by a complete DNA sample appear to be very convincing statistics, but we should remember not to be blindsided by these very big or small numbers in the courtroom. We should always be careful to consider the circumstances that led to their generation and remember that, without proper interpretation, simply quoting an extremely small number out of context does not in itself demonstrate the guilt or innocence of a suspect.

The "one in 73 million" figure concocted by Meadow in Sally Clark's case is one such cautionary example. The combination of erroneous independence assumptions (assuming that having had one baby die due to SIDS doesn't alter the probability of having a second baby die from SIDS) and ecological fallacies (pigeonholing the Clarks into a lower-risk category based on some cherry-picked demographic details) made the figure far smaller than it should have been. To compound these problems, the figure was also so presented that any reasonable jury might have assumed that one in 73 million was the probability of Sally's innocence, rather than its being the probability of one possible alternative explanation for the infants' deaths—the prosecutor's fallacy. Indeed, a jury did find her guilty, based, in no small part, on Meadow's presentation of his incorrect figure.

If we should be careful of being too readily convinced of someone's guilt by extremely small probabilities, we should not simply accept the refutation of these figures as evidence of someone's innocence. Andrew Deen fell prey to the prosecutor's fallacy, making the probability of his guilt, based on the DNA evidence alone, seem much more likely than it was. At his appeal, Deen's defense argued for a revised figure of one in twenty-five hundred for the probability of a DNA match, making him one of thousands of potentially matching suspects in the vicinity of the crime. One might argue that this makes the DNA evidence effectively worthless. This argument, however, is equally erroneous and is known as the defense attorney's fallacy. The DNA evidence should not be discarded, but assimilated alongside the other evidence implicating or exonerating a suspect. Deen's conviction was overturned in part because of the misleading effect the prosecutor's fallacy had on the jury. At his retrial, however, Deen pled guilty and was convicted of rape.

In the same way, Schneps and Colmez present a compelling mathematical argument that by denying a retest of the DNA, Judge Hellmann, presiding at Amanda Knox's appeal, may have helped secure her freedom. In 2013, Knox's appeal acquittal was quashed, and a judge ordered that the second DNA sample be retested. The DNA was found to belong to Knox herself. At her final appeal, in 2015, judges heard evidence that the collection and examination of the knife had been severely compromised. Errors ranged from the knife's being collected and stored in an unsealed envelope and then an unsterilized cardboard box, to police officers not wearing the correct protective clothing, and even to one officer's having been at Kercher's apartment before handling the knife later in the day. Contamination in the lab was also difficult to rule out, with at least twenty of Kercher's samples having been tested in the lab before the alleged murder weapon was examined. If the original DNA found on the knife was there by contamination, no number of retests would change the fact that the DNA belonged to Kercher or would answer how it came to be on the knife. Indeed, had more of the contaminated DNA sample been available, retesting could erroneously have provided more misplaced confidence of Knox's guilt.

By getting hung up on the details of a neat mathematical argument, a complex computation, or a memorable figure, we often neglect to ask the most pertinent question: Is the calculation in question even relevant?

In Sally Clark's case, the statistic that most influenced the jurors was Meadow's estimate of the likelihood of two SIDS deaths in a single family. Upon closer analysis we might question why this figure was even calculated. No one at the trial was arguing that both of the Clarks' children died of SIDS. At the time of Christopher's death, the pathologist conducting the postmortem certified that Christopher had died of a lower respiratory tract infection. This is not the same as a diagnosis of SIDS, which is effectively the diagnosis reached when everything else has been ruled out. The defense claimed natural causes, the prosecution claimed murder, but no one suggested that SIDS was the cause of both infants' deaths. Meadow's figure purporting to describe the probability of two SIDS deaths in the same family had no business being in the courtroom. Yet this figure seems to have been a significant factor in the minds of the jurors when they concluded that Sally was guilty of murdering her two infant sons.

At her second appeal, in January 2003, Sally's lawyers presented new evidence that had been discovered since her original conviction. The evidence from the postmortem of Sally's second son, Harry, clearly indicated the presence of the bacteria *Staphylococcus aureus* in his cerebrospinal fluid. According to experts, this infection was extremely likely to have led to some form of bacterial meningitis that caused Harry's death. Although the new microbiology evidence was enough for Sally's conviction to be overturned, the appeal judges stated that the misuse of statistics in the original trial would have been enough to uphold the appeal.

On January 29, 2003, Sally was set free. She returned to Steve and her third son, who was by then four years old. In a statement given upon her release, she talked of being allowed finally to grieve for the deaths of her babies, the importance of returning to her husband, of being a mother to her little boy, and of becoming a "proper family again." Despite her overwhelming delight at being reunited with her family, even this reprieve was not enough to make up for the years she had spent wrongfully incarcerated, blamed for killing two of the people she loved most. In March 2007, she was found dead at her home from alcohol poisoning, never having fully recovered from the effects of her wrongful conviction.

We can extend the lessons learned in the courtroom to other areas of our lives. As we will see in the next chapter, it is prudent to bring a questioning attitude to the figures that catch our eyes in newspaper headlines, the claims pushed at us in ads, or the dubious rumors we catch from our friends and colleagues. In any area where someone has a vested interest in manipulating the figures, which is almost everywhere figures occur, we should treat claims skeptically and ask for more explanation. Anyone who is confident in the veracity of their figures will be happy to provide it. Math and statistics can be difficult to understand, even for trained mathematicians; this is why we have experts in those areas. If need be, ask for help from a professional, a Poincaré, who can lend an expert opinion. Any mathematician worth their salt will be happy to oblige. Even more important, before a mathematical smoke screen is whipped up in front of us, we must vigorously question whether mathematics is even an appropriate tool to use.

With the increasing prevalence of quantifiable forms of evidence, mathematical arguments have an irreplaceable role in some parts of our modern justice system, but in the wrong hands mathematics can act as a tool that impedes justice, costing innocent people their livelihoods and, in extreme cases, their lives as well.

CHAPTER 4

DON'T BELIEVE THE TRUTH

Debunking Media Statistics

D *on't Believe the Truth* was the title of Mancunian rock band Oasis's sixth album. Growing up in Manchester in the 1990s, I was mad for the band. I'd been to see them at a number of venues around the city, and just after this album came out, in 2005, I went to see them play again, at the City of Manchester Stadium, home of my beloved Manchester City Football Club. As a teenager I used to go to gigs fairly regularly in a range of venues around Manchester: the Apollo, Night and Day, the Roadhouse, and for bigger bands the Manchester Arena.

By 2017, Oasis had long since broken up and I hadn't lived in Manchester or been to a gig there for over ten years, but many of the music venues I used to frequent were still going strong. On May 22 that year, at around half past ten in the evening, an Ariana Grande concert had just finished at the Manchester Arena. The audience, many of whom were teenagers or younger, were streaming into the foyer to meet their waiting parents. Motionless, in the middle of the crowd, stood twenty-three-year-old Salman Abedi. On his shoulders he carried a rucksack filled with the nuts and bolts that enveloped his homemade bomb. At 10:31 he detonated it. He killed twenty-two innocent victims and injured hundreds more. It was the worst terrorist attack on UK soil since the 2005 bombings, which targeted the London transport network, killing fifty-two members of the public.

At the time of the concert attack I was not in Manchester; I was not even in the country. I was visiting Mexico City for work. Because of the six-hour time delay, I watched reports of the attack come in one after the other as my afternoon progressed and most of the UK slept, as yet unaware. Despite being over five thousand miles away, having traversed that very foyer after a gig myself, I felt somehow more connected to the incident, more shocked and more appalled than by many recent terrorist incidents. Over the next few days I read as much as I could about the attack and how the people of my hometown had reacted. One article, in the UK tabloid *Daily Star*, particularly caught my attention. It was entitled "'Dates Matter to Jihadis' Manchester Arena Attack on Lee Rigby Anniversary." The article's author highlighted a tweet by Sebastian Gorka, then deputy assistant to US president Donald Trump, which read, "Manchester explosion happens on 4th anniversary of the public murder of Fusilier Lee Rigby. Dates matter to Jihadi terrorists."

Gorka had noticed a coincidence between the dates of two Islamist terrorist attacks. The first, on May 22, 2013, was a butcher knife attack on a British Army soldier by two Christianity-to-Islam converts of Nigerian descent. The second, on May 22, 2017, was a suicide bombing on a nonpolitical target by a lifelong Muslim of Libyan ancestry. Gorka was suggesting, in his tweet, that the Manchester Arena attack was meticulously planned to be carried out on the anniversary of Lee Rigby's murder. If this was true, it would lend credibility to the idea that Islamist terrorists are a well-organized and coherent group, capable of striking at will on a chosen date. This, however, is somewhat at odds with the "lone wolf" picture that has since been painted of Abedi.

Organization and order in a terrorist group seems to make them more threatening than if attacks are seen to be carried out at random with no central control or coherence. The purpose of Gorka's tweet seemed to be to heighten the fear of Islamist terrorism, perhaps with the aim of supporting President Trump's embattled executive order "Protecting the Nation from Foreign Terrorist Entry into the United States," which banned many Muslims from traveling to the country and which was, at the time, undergoing several legal challenges. But is this really the case? I wondered. Should we believe Gorka's assertion, given credence by the *Daily Star*? Isn't this the sort of unfounded, hyped-up rhetoric that serves the terrorists' purposes perfectly? How likely is it, I wondered, that two terrorist incidents would happen on the same day of the year purely by chance?

We are constantly bombarded by numbers and figures in what we read, what we watch, and what we hear. Large-cohort studies into the ways in which twenty-first-century lifestyles impact our health, for example, are accruing faster than ever. Simultaneously, there is a concomitant increase in the numerical skills required to interpret their findings. In many cases there is no hidden agenda, the statistics are just difficult to interpret. However, for many reasons it might benefit one party or another to put a spin on a particular finding.

In the era of fake news, it's difficult to know whom to trust. Believe it or not, most mainstream media outlets base most of their stories on facts. Truthfulness and accuracy are near the top (if not at the top) of the list on almost all codes of journalistic ethics and integrity. In addition to moral obligations to tell the truth, libel cases can be extremely damaging and expensive, providing a financial incentive to get the facts right.

Many media organizations differ in their reporting of the facts, however, in their slants on stories. When, for example, President Trump's tax reform bill (its title, Tax Cuts and Jobs Act, is not without spin itself) was passed in December 2017, journalist Ed Henry on Fox reported it as "a major victory" and a "desperately needed win for the president." Lawrence O'Donnell on MSNBC, however, referred to Republican senators who voted for the bill as providing "the ugliest display of pigs at the trough that I have ever seen in Congress." Jake Tapper on CNN led with the question "Has there ever been a piece of major legislation passed by the Congress with less [popular] support?"

You will have had no trouble detecting the different verbal bias applied to the above story and making inferences about the political agendas promoted by the three news outlets. Partisanship is easy to detect through people's words. Numbers, on the other hand, are easier to spin surreptitiously. Statistics can be cherry-picked to present a particular angle on a story. Other figures are ignored altogether and misrepresentative stories created purely by omission. Sometimes the studies themselves are unreliable. Small, unrepresentative, or biased samples, in conjunction with leading questions and selective reporting, can all make for unreliable statistics. More subtle still are the statistics used out of context so that we have no way to judge whether, for example, a 300 percent increase in cases of a disease represents an increase from one patient to four or from half a million patients to 2 million. Context is important. It's not that these different interpretations of numbers are lies — each one is a small piece of the true story on which someone has shone a light from a preferred direction — it's just that they are not the whole truth. We are left to try to piece together the true story behind the hyperbole.

In this chapter, we will analyze and demystify the tricks, traps, and transformations in newspaper headlines, advertisements, and political sound bites. We will expose similar mathematical manipulations employed in places where we might expect better: in patient-advice publications and even in scientific articles. We will provide simple ways of recognizing when we are not being told the whole story and tools to help us reverse the spin applied to a statistic, as we try to find out when we should believe the "truth."

The Birthday Problem

The most subtle, and often effective, mathematical misdirections are the ones in which it doesn't even appear that a number is at play. By stating "Dates matter to Jihadi terrorists," Gorka was implicitly asking us to evaluate the probability that two terrorist incidents could have occurred on the same day by chance, making it clear that he didn't think that this was likely. The way to find out the real answer lies in a mathematical thought experiment known as the birthday problem.

The birthday problem asks, "How many people do you need to have at a gathering before the probability of at least two people sharing a birthday rises above 50 percent?" Typically, when first posed this problem, people plump for a number such as 180, which is roughly half the number of days in the year. This is because we tend to put ourselves in the room and think about the probability of someone else matching our own birthday. One hundred and eighty is, in fact, way, way too many. Making the reasonable assumption that birthdays are roughly evenly distributed throughout the year, the answer is just twenty-three people. This is because we are not concerned about the particular day on which the birthday falls, just that there is a match.

To gain an insight into why the number required is so low, we can start by considering the number of pairs of people in the room—pairs of birthdays falling on the same day, after all, being what the question is about. To calculate the number of pairs with 23 people in the room, imagine lining them all up and asking them to shake hands with each other. The first person shakes hands with the other twenty-two people, the second with the twenty-one people she has not shaken hands with yet, the third with twenty others, and so on. Finally, the penultimate person shakes hands with the last and we are left to add up $22 + 21 + 20 + \ldots + 1$. This is arduous, though easy enough for twenty-three people, but borders on tedious when the number of people in the room gets above fifty. Sums such as this—of consecutive whole numbers, starting from one—are called triangular numbers, since we can lay out these numbers of objects in orderly triangular arrays, as we have done in figure 15 (page 114). Fortunately, there is a neat formula for triangular numbers. For a general number of people, N, in the room, the number of handshakes is given by $N \times (N - 1)/2$. For 23 people, this gives $23 \times 22/2$ or 253 pairs. Perhaps it's not surprising then that the probability of at least one pair of people with the same birthday rises to above 50 percent with so many pairs of people in the room.

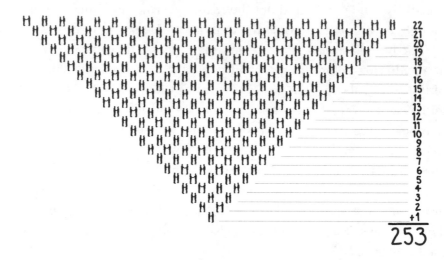

FIGURE 15: The number of handshakes between 23 people. The first person shakes hands with 22 others, the second with 21, and so on until the penultimate person is left with only one person to shake hands with. The total number of handshakes between 23 people is the sum of the first 22 whole numbers. The formula for the triangular numbers tells us that there are 253 pairs of people with just 23 people in the room.

To put a figure on this probability, it's easier to think first about the probability that nobody shares a birthday. This is exactly the same mathematical technique that we employed in chapter 2 when calculating how many mammograms a woman might undergo before the probability of her receiving a false-positive diagnosis increased to above one-half. With any single pair of people we can easily find the probability they don't share a birthday. The first person can have their birthday on any of the 365 days of the year, and the second on any of the remaining 364 days. So the probability of a single pair of people not sharing a birthday is pretty close to certain, at 364/365 (or 99.73 percent). However, with 253 pairs of people, and we are interested in finding out the probability that none of them share a birthday, we need each of the other 252 pairs to also have distinct birthdays. If all of these pairings were independent of each other, the probability that none of the 253 pairs of people share a birthday would be given by the probability that one pair don't share a birthday, 364/365, multiplied by itself 252 times or $(364/365)^{253}$. Although 364/365 is quite close to 1, when multiplied by itself hundreds of times, the probability of no matching pairs of birthdays turns out to be 0.4995, just under one-half. Since nobody sharing a birthday or two or more people sharing a birthday are the only two possibilities (in mathematical parlance they are "collectively exhaustive"), the probabilities of the two events must sum to one. Therefore, the probability that two or more people share a birthday is 0.5005, just more than one-half.

In reality, not all the pairs of birthdays will be independent of each other. If person A shares a birthday with person B, and person B shares a birthday with person C, then we know something about the pairing A-C: they must also share a birthday—they are no longer independent. If they were independent, they would have only a 1/365 chance of sharing a birthday. The exact calculation of the probability of a match, taking these dependencies into account, is only slightly more involved than when we assumed independence in the previous paragraph. It relies on thinking about adding people to the room one at a time. For two people, we established that the probability of not sharing a birthday is 364/365. Adding a third person to the mix, they can have their birthday on any of the remaining 363 days of the year if they are not to share a birthday with either of the others. So the probability of three people not sharing a birthday is $(364/365) \times (363/365)$. A fourth person can only have their birthday on one of the 362 remaining days, so the probability of all four not sharing a birthday drops slightly to $(364/365) \times (363/365) \times (362/365)$. The pattern continues until we add the twenty-third person to the party. They can have their birthday on any of the remaining 343 days of the year. The probability that twenty-three people don't share a birthday is given by the protracted multiplication

$$\frac{364}{365} \times \frac{363}{365} \times \frac{362}{365} \times \cdots \times \frac{343}{365} .$$

This expression tells us that the exact probability that two people in a group of twenty-three don't share a birthday (accounting for possible dependencies) is 0.4927, just under one-half. Using the idea of collective exhaustion again (that either no shared birthdays or at least one shared birthday are the only two options), the only other possibility—that at least two people do share a birthday—has a probability just over one-half, at 0.5073. By the time seventy people are in the group, there are 2,415 pairs of people. The exact calculation tells us that the probability of a match is overwhelmingly likely, at 0.999. Figure 16 shows how the probability of two events falling on the same day of the year changes as the number of independent events we consider increases from one to a hundred.

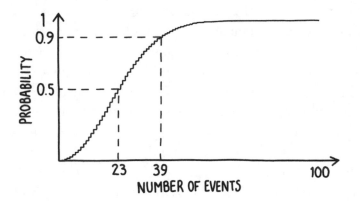

FIGURE 16: The probability of two or more events falling on the same day increases with the number of events. When there are 23 events, the probability of a match is just above 1/2. When there are 39 independent events the probability of at least 2 of them falling on the same day rises to nearly 0.9.

I used the surprising results of the birthday problem to impress my literary agent when we met up for the first time to discuss writing this book. I bet him the next round of drinks that I would be able to find two people, in the relatively quiet pub, who shared a birthday. After a quick scan of the room, he readily took me on and offered to buy the next *two* rounds if I could find such a pair, so unlikely did he think the prospect of a match. Twenty minutes and a lot of baffled looks and superficial explanations later ("It's okay," a slightly worse-for-wear version of myself told the people I accosted, "I'm a mathematician"), I had found my pair of birthday sharers and the drinks were on Chris. It probably wasn't quite fair of me; I'd already done my own head count when I went to the bar for the previous round and counted about forty customers. With that number of customers, I had only an 11 percent chance of losing the bet. I should have been staking the next *two* rounds against Chris's one, not the other way around. More than just a facile math hustle to exploit unsuspecting victims in the pub, though, the high probability of a match for surprisingly low numbers of events has some far deeper implications. In particular, it can help us test Gorka's implication of jihadis' ability to strike at will.

Over the five-year period between April 2013 and April 2018, at least thirty-nine terrorist attacks against Western nations (European Union, North American, or Australian) were committed by Islamist terrorists. At first glance it seems unlikely that two of these would fall on the same day if they simply occurred at random throughout the year. However, with 741 possible pairs of events, the probability of two falling on the same day is likely indeed, at around 88 percent, as captured in figure 16. With this high probability we should be extremely surprised if two of these attacks did not fall on the same day. This says nothing about the likelihood of future terrorist attacks, but it seems Gorka was giving more credit to the organizational skills of Islamist terrorists than they deserve.

The same birthday-problem reasoning tells us that we have to be careful how we interpret the DNA evidence that is so instrumental in many modern criminal trials (as exemplified in the previous chapter). In 2001, while searching through Arizona's offender DNA database of 65,493 samples, a scientist discovered a partial match between two unrelated profiles. Nine out of thirteen loci matched between the samples. To put that in perspective, for two given unrelated individuals we would only expect a match of this caliber roughly once in every 31 million profiles sampled. This shock finding prompted a search for more possible matches. By the time all the profiles in the database were compared, 122 pairs of profiles matching at nine or more loci had been found.

Based on this study and now doubting the uniqueness of DNA identifiers, lawyers across the United States argued for similar comparisons to be made in other DNA databases, including the national DNA database, containing 11 million samples. If 122 matches had turned up in a database as small as 65,000 people, could DNA be relied upon to uniquely identify suspects in a country with a population of 300 million? Were the probabilities associated with DNA profiles incorrect and therefore risking the accuracy of DNA-based convictions across the nation? Some lawyers believed so and even submitted the Arizona findings as evidence to cast doubt on the reliability of the DNA evidence in their defendants' trials.

Using the formula for the triangular numbers, we can work out that comparing each of the 65,493 samples in the Arizona database with every other gives a total of over 2 billion unique pairs of samples. With a probability of one match per 31 million pairs of unrelated profiles, we should expect 68 partial (i.e., matching at nine loci) matches. The difference between the expected 68 matches and the 122 that were found might easily be explained by the profiles of close relatives in the database. These profiles are significantly more likely to throw up a partial match than those of unrelated individuals. Rather than shaking our confidence in DNA evidence, in the light of the insight gained from triangular numbers, the database findings agree nicely with the mathematics.

Authority Figures

In the original *Daily Star* article highlighting the coincidence in the dates of the murder of Fusilier Lee Rigby and the Manchester Arena attack, the probability we needed to evaluate so as to assess Gorka's claim was hidden out of sight. This is in direct contrast to the way in which most advertisers use figures. If flattering-enough figures can be found, then they are generally presented prominently. Advertisers know that numbers are widely perceived as being indisputable facts. Adding a figure to an ad can be extremely persuasive and lend power to the promoter's argument.

The apparent objectivity of statistics seems to say, "Don't just trust what we're saying, trust this piece of indisputable evidence."

Between 2009 and 2013, L'Oréal advertised and sold the Lancôme Génifique line of "antiaging" products. Alongside the usual advertising pseudoscience ("Youth is in your genes. Reactivate it"; "Now, boost genes' activity and stimulate the production of 'youth proteins'") was a bar graph purporting to show that 85 percent of consumers had "perfectly luminous" skin, 82 percent had "astonishingly even" skin, 91 percent had "cushiony soft" skin, and 82 percent found skin's "overall appearance improved" after just seven days. The hopelessly nebulous description of the improvements aside, these figures sound extremely impressive, a ringing endorsement of the product.

Burrow down a little further into the study behind the figures, however, and we find a quite different story. Women who took part in the study were asked to apply Génifique twice a day and consider how they felt about statements including "Skin appears more radiant/luminous"; "Skin tone/complexion appears more even"; and "Skin feels softer." They were then asked to rate their agreement with the statements on a 9-point scale, whose extremes ranged from 1, "disagree completely," to 9, "agree completely." The subjects were not asked to rate the degree of radiance, softness, or evenness of their skin; just how much they agreed or disagreed that there was an improvement. They were certainly not asked to provide adverbs such as "perfectly" or "astonishingly."

The results of the survey showed that, although 82 percent of women did agree (giving a score between 6 and 9 on the 9-point scale) that their skin appeared more even after seven days, less than 30 percent "agree[d] completely." Similarly, although 85 percent agreed that their skin looked more radiant/luminous, only 35.5 percent agreed completely. L'Oréal had presented the results of their own survey in a way that made them appear more impressive than they were.

Perhaps of more concern was the size of the study. With just thirty-four participants, it's hard to be sure that the results are reliable because of an effect known as small sample fluctuations. Small sample sizes will typically show greater deviations from the true population mean than large samples. To illustrate this, imagine I have a fair coin—one that comes down heads 50 percent of the time and tails 50 percent of the time. For some reason, I want to convince people that the coin is biased in favor of tails. Let's say they will be convinced if I can show them that the coin comes up tails at least 75 percent of the time. How do my chances of winning them over change as the sample size—the number of flips of the coin—increases?

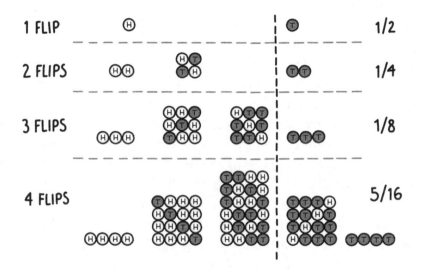

FIGURE 17: All possible heads and tails combinations that can arise for one, two, three, and four coin flips. The combinations to the right of the vertical line are all at least 75 percent tails.

I might try to get away with just flipping the coin once. If it comes up tails, I'm happy; one tail in one flip exceeds the 75 percent threshold. This happens on half of the occasions that I flip the coin once. A single flip gives me the best chance of convincing someone that the coin is biased, but people would be right to object that they need to see more data to be convinced, and to ask me to flip the coin again. With two flips I need to achieve two tails to convince people that the coin is biased. One tail and one head won't cut it as the number of tails is only 50 percent. As we can see in figure 17, two tails is just one of the four equally likely outcomes from two flips of an unbiased coin, so I win over only a quarter of the people I try to convince. The probability of seeing at least 75 percent tails diminishes rapidly as the size of the sample is increased, as seen in figure 18. By the time I'm asked to increase the sample size to one hundred flips, my chances of convincing someone that the coin is biased drop to 0.00000009.

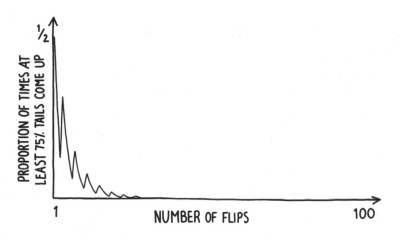

FIGURE 18: The chances of convincing someone that a genuinely fair coin is biased in favor of tails diminishes rapidly as the number of flips increases.

As the sample size increases, the variation around the mean (in this case the mean would be 50 percent tails) decreases: it becomes harder and harder to convince someone of something that isn't true. That's why, with only thirty-four participants in the study, we should be skeptical about the reliability of the results presented in L'Oréal's ad.

Typically, ads with small sample sizes report their findings in percentages (82 percent had astonishingly even skin) rather than ratios (twenty-eight of thirty-four had astonishingly even skin) to hide the embarrassingly small sample sizes. The telltale sign of the small sample size, however, is if, as in the Génifique ad, you find two percentages the same (82 percent also found the skin's overall appearance improved). You have relatively few options to choose from in a small sample size if you want your audience to believe that your product is good, but not too good (figures between 95 percent and 100 percent, for example, might look suspicious). With a larger sample size it's far less likely that exactly the same number of people will give positive answers to two different questions.

In 2014, the Federal Trade Commission (FTC) wrote to L'Oréal charging them with deceptive advertising over the Génifique range. The FTC claimed that the numbers in the ads' charts were "false or misleading" and not proven by scientific studies. In response, L'Oréal agreed to stop "making claims about these products that misrepresent the results of any test or study."

As well as small sample bias, the Génifique study might also have suffered from sampling biases such as "voluntary response bias" or "selection bias." If L'Oréal recruited participants for the study by placing an ad on their website, for example, then they would likely recruit women who were already susceptible to the perceived benefits of the product and likely to give it a good review (voluntary response bias). Alternatively, L'Oréal might have handpicked the women specifically because they had given good reviews on L'Oréal products in the past (selection bias).

Favorable figures for a study, poll, or political sound bite can be arrived at in other, even more dubious ways. If the first study of thirty-four participants doesn't come back with a favorable result, why not do another? Sooner or later the large variation will give you the impressive responses you need. Alternatively, why not just run a larger trial and cherry-pick the participants who give you the best responses. This is known as data manipulation or, less technically, "fudging the data." A common example of this phenomenon is reporting bias. Scientists who investigate pseudoscientific phenomena such as alternative medicine or extrasensory perception (psychic abilities) often bemoan what they perceive as reporting bias among investigators sympathetic to the cause. Unscrupulous researchers present only the "positive results" (participants who report a benefit from a treatment, for example, or runs of a "psychic" choosing the correct color of the next card in a shuffled deck), while the majority of "negative results" are discarded, making findings seem more favorable than they actually are. When two or more types of biases combine, they can give wildly different results to those expected in an unbiased sample, as the editors of the *Literary Digest* magazine discovered.

Hard to Digest

In the lead-up to the 1936 election to determine the president of the United States, the editors of the well-respected monthly magazine *Literary Digest* conducted a poll to predict the winner. The candidates were the incumbent president, Franklin D. Roosevelt, and his Republican challenger, Alf Landon. The *Digest* had a proud history of correctly predicting the next president, going all the way back to 1916. Four years earlier, in 1932, they had predicted Roosevelt's victory margin to within a percentage point. In 1936, their poll was to be as ambitious and expensive as any poll ever previously undertaken. The *Digest* created a list of around 10 million names (roughly one-quarter of the voting population) based on automobile-registration records and names in telephone directories. In August, they sent out straw polls to everyone they had identified and trumpeted in the magazine, "If past experience is a criterion, the country will know to within a fraction of 1 percent the actual popular vote of forty million [voters]."

By October 31, over 2.4 million votes had been returned and counted. The *Digest* announced their result: "Landon, 1,293,669; Roosevelt, 972,897." The *Digest* had Landon winning by a wide margin: 55 percent to 41 percent of the popular vote (with a third candidate, William Lemke, polling 4 percent) and taking 370 of the 531 electoral votes. Only four days later, when the genuine election results were announced, the editors of the *Digest* were shocked to find that Roosevelt was reelected to the White House. It wasn't a narrow victory either, but a landslide. Roosevelt won 60.8 percent of the popular vote—the largest share since 1820. He took 523 of the electoral votes to Landon's 8. The *Digest* was out by nearly 20 percentage points in their prediction of the popular vote. We might expect a large variation in the results with a small sample size, but the *Literary Digest* had polled 2.4 million people. With such a large sample, how did they get it so wrong?

The answer is sampling bias. The first issue the poll suffered from was selection bias. In 1936, America was still in the grip of the Great Depression. Those people who owned cars and telephones were likely among the better off in society. Consequently, the list that the *Digest* compiled was skewed toward the upper- and middle-class voters, among whom, leaning further toward the right in their political opinions, support for Roosevelt was less strong. Many of the poorer people, who comprised Roosevelt's core support, were completely unaccounted for in the poll.

Perhaps even more significant for the results of the poll was "nonresponse bias." Of the 10 million names on the original list, fewer than a quarter responded. The survey was no longer sampling the population it originally intended to. Even if the initial demographic selected had been representative of the population as a whole (which it wasn't), the people who responded to the survey tended to have different political attitudes from the people who didn't. The typically wealthier and better-educated people who did answer tended to be supporters of Landon, rather than Roosevelt. Together, these two sampling biases combined to give embarrassingly incorrect results, leaving the *Digest* a laughingstock.

That same year, using just forty-five hundred participants, *Fortune* magazine predicted the margin of Roosevelt's victory to within 1 percent. The *Literary Digest* did not come out well in comparison. The dent that their previously impeccable credibility sustained because of these results is cited as a significant factor in hastening the magazine's demise less than two years later.

Do the Math

While political pollsters have found they need to be increasingly statistically aware to get accurate results, politicians themselves are finding that they can get away with more statistical manipulation, misappropriation, and malpractice than ever before. When running for the Republican nomination in November 2015, Donald Trump tweeted an image with the following statistics:

Blacks killed by whites — 2 percent
Blacks killed by police — 1 percent
Whites killed by police — 3 percent
Whites killed by whites — 16 percent
Whites killed by blacks — 81 percent
Blacks killed by blacks — 97 percent

The source of the figures was attributed to the "Crime Statistics Bureau — San Francisco." As it turns out, the Crime Statistics Bureau doesn't exist, and the statistics are wildly off the mark. Some real comparative statistics for 2015 (with raw figures given in table 10, page 127) from the FBI are:

Blacks killed by whites — 9 percent
Whites killed by whites — 81 percent
Whites killed by blacks — 16 percent
Blacks killed by blacks — 89 percent

Evidently, Trump's tweet massively overplayed the number of homicides committed by black people, effectively transposing the figures for "white-on-white" and "black-on-white" murders. Nevertheless it was retweeted over seven thousand times and "liked" over nine thousand times. This is a classic example of confirmation bias. People retweeted the false message because it came from a source they respected and it chimed with their preexisting prejudices. They didn't stop to check if it was true, and neither did Trump. When he was questioned by journalist Bill O'Reilly on Fox News about his motivation for disseminating the image, after first claiming, in typical hyperbolic style, "I'm probably the least racist person on earth," he followed up with ". . . Am I gonna check every statistic?"

Trump's tweet in 2015 came at the height of the US national debate about police brutality, particularly brutality targeting black victims.

Such cases, including most notably the deaths of the unarmed black teenagers Trayvon Martin and Michael Brown, were the catalyst for the formation and rapid expansion of the Black Lives Matter movement. Between 2014 and 2016, Black Lives Matter held mass protests, including marches and sit-ins, across the United States. By September 2016 the movement had chapters in the UK, whose protests drew the ire of right-leaning journalist Rod Liddle. A mathematically oriented blog post drew my attention to Liddle's comments in the British tabloid newspaper the *Sun* on the foundation of the original Black Lives Matter movement in the United States:

> It was set up to protest about American cops shooting black suspects instead of just arresting them. There's no doubt US cops are a bit trigger-happy. And maybe especially when a black suspect hoves into view. There's also no doubt whatsoever that the greatest danger to black people in the USA is . . . er . . . other black people. Black-on-black murders average more than 4000 each year. The number of black men killed by US cops—rightly or wrongly—is little more than 100 each year. Go on, do the math.

So here it is—I did the math.

Let's consider the statistics for 2015, the latest full calendar year for which Liddle could have accessed data. According to FBI statistics (summarized in table 10) 3,167 white people and 2,664 black people were murdered in 2015. Of the homicides in which the victim was white, 2,574 (81.3 percent) were perpetrated by white offenders and 500 (15.8 percent) by black offenders. Of the homicides in which the victim was black, 229 (8.6 percent) were perpetrated by white offenders, and 2,380 (89.3 percent) by black offenders. So Liddle's claim of 4,000 "black-on-black" homicides a year is a significant exaggeration: by approximately 70 percent. Given that black people comprised just 12.6 percent of the US population in 2015 and white people 73.6 percent, it is alarming that black individuals make up 45.6 percent of the homicide victims.

RACE/ETHNICITY OF VICTIM	TOTAL	RACE/ETHNICITY OF OFFENDER	
		WHITE	BLACK
WHITE	3,167	2,574 (81.3%)	500 (15.8%)
BLACK	2,664	229 (8.6%)	2,380 (89.3%)

TABLE 10: Homicide statistics for 2015 broken down by the race/ethnicity of the victim and perpetrator. The disparities between the total column and the sum of white and black victim columns are due to cases in which the ethnicity of the victim is different or unknown.

Although a far more prominently debated issue, figures for the number of people killed by police are harder to obtain. The fatal shooting of black teenager Michael Brown by white police officer Darren Wilson, and the subsequent protests that took hold of Ferguson, Missouri, marked a tipping point for the Black Lives Matter movement. The protests also shone a spotlight on the FBI's "annual count of homicides by police." The FBI was found to be recording fewer than half of all killings by police in the United States. In response, in 2014, the *Guardian* began its campaign "The Counted" to compile more accurate figures. So successful was the project that, in October 2015, FBI director James Comey called it "embarrassing and ridiculous" that the *Guardian* had better data on civilian deaths at the hands of the US police than the FBI.

The *Guardian*'s figures show that, of the 1,146 people "rightly or wrongly" (to echo Liddle) killed by police in 2015, 307 (26.8 percent) were black and 584 (51.0 percent) were white (while the remaining victims were of different or undetermined ethnicities). Again, Liddle's figure is way off the mark. His suggestion of 100 police killings of black citizens a year is less than a third of the true value.

If Liddle was trying to answer the question "If a black person in the United States is killed, is it more likely to have been another black person or a police officer who killed them?," then using the correct figures, it's clear that black people kill almost eight times (2,380 versus 307) more black people than police do. But this question seems disingenuous. Would you believe that dogs are more murderous than bears if I told you that, in 2017, dogs killed forty US citizens while bears killed just two? Of course not. Dogs are not inherently more dangerous than bears, there are just more of them in the United States. To put it another way, if you were to be left alone in a room with a bear or a dog, which would you prefer it to be? I don't know about you, but I'd probably opt for the dog.

For the same reason, given that there are over 40.2 million black US citizens and only 635,781 full-time "law enforcement officers" (those who carry a firearm and a badge), it's not surprising that more killings are perpetrated by black people than law enforcement officers. A more appropriate question for Liddle to ask might have been, "If a black US citizen comes across someone while out walking alone, who should they be more scared will kill them: another black person, or a law enforcement officer?"

To find out the answer we need to compare the per capita rates of black-victim killings perpetrated by black people and by police officers. We find the per capita rates, as presented in table 11, by dividing the total number of black victims killed by a particular group (black people or police officers) by the size of the group. Black people were responsible for 2,380 killings of other black people in 2015, but with over 40.2 million black US citizens, the per capita rate is relatively small—around one in seventeen thousand. Police officers were "rightly or wrongly" responsible for killing 307 black people in 2015. With 635,781 police officers, this amounts to a per capita killing rate that is just below one killing per two thousand police officers—over eight times higher than the rate for black US citizens. It seems that a black person walking down the street should be more alarmed to see a police officer approaching than another black person.

KILLER	NUMBER OF BLACK-VICTIM KILLINGS	SIZE OF POPULATION	PER CAPITA KILLING RATE
BLACK CITIZENS	2,380	40,241.818	1/16,908
LAW ENFORCEMENT OFFICERS	307	635,781	1/2,071

TABLE 11: The number of killings in which a black citizen was the victim, stratified by whether the killing was committed by another black person or a law enforcement officer. The sizes of the two populations are also presented and used to work out the per capita killing rate.

Of course we have not accounted for the fact that encounters with the police are often confrontational, and US police are typically armed. It's perhaps not surprising that those authorized to wield lethal force do so more frequently than the general population at large. By exactly the same mathematics, we can show that white people should also be more scared of law enforcement officers (per capita white killing rate of one per thousand officers) than other white people (per capita white killing rate of one per ninety thousand white people), despite more white people killing other white people than police officers killing white people. That police officers have twice as high a per capita rate of killing white people than black people is because the country has more white people. That the rate is only twice as high, given that the United States has almost six times as many white people as black people, is perhaps unsettling.

So while Liddle's statistics are incorrect, perhaps more important, by asking "Who kills the most?" rather than "Who is killed the most?," his *Sun* article diverts attention from a statistic that is at the heart of the Black Lives Matter movement: that the 12.6 percent of the population who are black account for 26.8 percent of police killings, while the 73.6 percent who are white account for just 51.0 percent. Are there hidden links (the type of lurking variables we met in the previous chapter explaining the supposed benefits of smoking to low-birth-weight babies) that might explain this disparity? Almost certainly there are. For example, poorer people are more likely to commit crime, and in the United States, black people are more likely to be poor. Whether these factors account for the huge overrepresentation of black people in the police homicide figures remains to be seen.

Careless Pork Costs Lives

Liddle's piece wasn't the first or the last time that the *Sun* newspaper was embroiled in statistical controversy. In 2009, under the, admittedly inspired, headline "Careless Pork Costs Lives," the *Sun* reported just one of many hundreds of results from a five-hundred-page study, by the

World Cancer Research Fund, on the effect of consuming fifty grams of processed meat per day. The tabloid newspaper shocked readers with the "fact" that eating a bacon sandwich every day would increase the risk of colorectal cancer by 20 percent.

But the figure was sensationalized. When stated in terms of "absolute risks"—the proportion of people exposed or unexposed to a particular risk factor (e.g., eating bacon sandwiches or not eating bacon sandwiches) who are expected to develop a given outcome (e.g., cancer) in each case—it turns out that fifty grams of processed meat per day increases the absolute lifetime risk of developing colorectal cancer from 5 percent to 6 percent. On the left of figure 19 (page 131) we consider the fates of two groups of one hundred individuals. Of one hundred people who eat a bacon sandwich every day, only one more of them will develop colorectal cancer than in a group of one hundred people who abstain.

Instead of using the more objective absolute risk, the *Sun* chose to focus on the "relative risk"—the risk of a particular outcome (e.g., developing cancer) for people exposed to a given risk factor (e.g., eating bacon sandwiches) as a proportion of the risk for the general population. If the relative risk is above 1, then an exposed individual is more likely to develop the disease when compared to someone without the exposure. If it is below 1, then the risk is decreased. On the right-hand side of figure 19, by neglecting the people who are not affected by the disease, the increase in relative risk (6/5 or, equivalently, 1.2) seems much more dramatic. Although it is true that the relative risk for those eating fifty grams of processed meat per day represents an increase of 20 percent, the absolute risk increased by only 1 percent. But an increased risk of 1 percent doesn't sell many papers. Sure enough, the article's headline was sufficiently inflammatory to spark the "Save our Bacon" media firestorm. Over the next few days, the furor over the figure saw scientists branded "health Nazis" who had declared a "war on bacon."

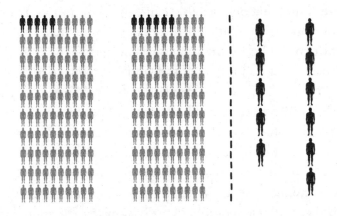

FIGURE 19: A comparison of the absolute figures (5 in 100 vs 6 in 100) (*left*) makes it appear that the increase in risk of eating 50 grams of processed meat per day is small. By focusing on the relatively small number of people who have the disease (*right*) the relative risk increase of 20 percent (1 in 5) seems very large.

Another attention-grabbing media trick is deliberately to change what we consider and accept to be the "normal" population. The most honest way to report relative risk is to present an increased or decreased risk for a particular subgroup in comparison to the background risk in the general population. Sometimes the disease risk levels of the largest subpopulation are used as the baseline, and any deviations in risk are reported relative to that population. When the disease is rare, the disease-free cohort makes up almost the whole population anyway, so the disease-free subpopulation risk is a good approximation to the general population risk. Consider reporting the risks of breast cancer for women with BRCA1 or BRCA2 genetic mutations, for example. It seems sensible to talk about the increase in absolute risk for the 0.2 percent of women with these mutations relative to the general population rather than a decreased risk for the 99.8 percent of women without these mutations. Unfortunately, this type of candid transparent reporting doesn't always make for the best headlines, so we see many of the big news outlets manipulating the way statistics are presented again and again to sell stories.

In a 2009 story headlined "Nine in 10 People Carry Gene Which Increases Chance of High Blood Pressure," the *Daily Telegraph* reported,

"Scientists found that one gene variant carried by almost 90 per cent of the population increased the chance of developing high blood pressure by 18 per cent." The figures actually reported in the journal *Nature Genetics* were that 10 percent of individuals possessed genetic variants that left them at a 15 percent lower risk than the 90 percent of the population with a different variant. The 18 percent figure didn't appear in the journal article. Although technically correct, the *Telegraph*'s story had mischievously changed the reference population to the smaller one—the 10 percent of people with the lowered risk. Since a 15 percent decrease from the reference value of 1 takes us to 0.85, the article's author recognized that the increase required to get back up to 1 is approximately 18 percent of this smaller figure. With one mathematical sleight of hand, the *Telegraph* not only increased the size of the relative risk but turned what might have been a good-news story for 10 percent of the population into a bad-news story for 90 percent of the population. The *Telegraph* was by no means alone in its manipulation of the figures—many other papers spun the story in the same dubious way to entice their audiences.

Often, after reading a sensationalist article, you will find you haven't been presented with the absolute risks—usually two small figures (certainly never more than 100 percent), one for those subject to the focal condition or intervention and the other for the remaining population. On other occasions it may be claimed that the risk is increased or decreased for more than half of the population. In these cases you should think carefully about whether to accept the article's argument. If you want to find out the truth behind the headlines, consider following up the story in a publication that gives you access to the absolute stats, or even find the original scientific article itself, which can increasingly be accessed online and for free.

A Different Frame of Mind

Newspapers are by no means alone in their dubious reporting of risks and probabilities. In the medical arena, in the communication of the risks of treatments or in the reporting of the efficacy of drugs and of their side effects, presenters can play more statistical games to advance an agenda. One simple way to suggest a particular interpretation is to frame figures in a positive or a negative light. In one study from 2010, participants were presented with a number of numerical statements about medical procedures and asked to rank the risk they associated with each on a scale from 1 (not risky at all) to 4 (very risky). In among them were the statements "Mr. Roe needs surgery: 9 in 1000 people die from this surgery" and "Mr. Smythe needs surgery: 991 in 1000 people survive this surgery." Take a second to think about whose shoes you'd rather be in: Mr. Roe's or Mr. Smythe's?

The two statements frame the same statistic in two contrasting ways— the first using mortality rates and the second survival rates. For participants with low numeracy skills, the positively framed statement about survival was perceived to be almost a whole point less risky on the 4-point scale. Even people with greater numeracy skills perceived the risk attached to the negatively framed statement as higher.

In medical trials one commonly sees positive outcomes reported in relative terms, to maximize their perceived benefit, while side effects are reported in absolute terms in an attempt to minimize the appearance of their risk. This practice is known as mismatched framing and was found to occur in roughly a third of the articles reporting the harms and benefits of medical treatments in three of the world's leading medical journals.

Perhaps even more worryingly, this phenomenon also appears prevalently in patient-advice literature. In the late 1990s, the US National Cancer Institute (NCI) created the Breast Cancer Risk Tool to educate and inform the public of their risks of the disease. Along with many other studies, the online app reported the results of a recent clinical trial on over thirteen thousand women at increased risk of breast cancer, in which the benefits and potential side effects of the drug tamoxifen were assessed. In the trial, the women were divided into two, roughly even, groups (known as the two "arms" of the trial). Women in the first arm were administered tamoxifen, while women in the second arm were given a placebo treatment as a control.

At the end of the five-year study, to assess the effect of the drug, the number of women with invasive breast cancer in each group were compared, as were the number of women with other types of cancer. In its Breast Cancer Risk Tool, the NCI reported the relative risk reduction: "Women [taking tamoxifen] had about 49 percent fewer diagnoses of invasive breast cancer." The big figure of 49 percent seems quite impressive. However, when it came to quantifying the possible side effects, an absolute risk was presented: "Annual rate of uterine cancer in the tamoxifen arm [of the trial] was 23 per 10,000 compared to 9.1 per 10,000 in the placebo arm." These tiny fractions seem to indicate that the risk of uterine cancer from the tamoxifen treatment doesn't change much at all. Consciously or unconsciously, while gathering data for their online risk-information tool, the NCI's researchers emphasized the benefits of tamoxifen for reducing breast cancer incidence while simultaneously minimizing the perception of the increased risk of uterine cancer. If these figures had been used to calculate a relative risk, to put the two statistics on a level playing field, it would have been reasonable to present a figure of 153 percent increased risk of uterine cancer to balance the 49 percent reduced risk of breast cancer.

Even in the abstract of the original article, the 49 percent figure for the reduction in breast cancer is used, but the increase in uterine cancer is presented as a relative risk ratio of 2.53. Using percentages instead of decimals to highlight perceived benefits is one of another family of tricks referred to as ratio bias. Our susceptibility to ratio bias has been confirmed in simple experiments in which blindfolded subjects are asked to choose at random a jelly bean from a tray. Drawing a red jelly bean represented a $1 win for the picker. When given the choice between a tray with nine white jelly beans and a single red one or a tray with 91 white jelly beans and nine red ones, the second tray is chosen more often, despite giving the smaller chance of picking out a winning sweet. Presumably this is because the higher number of red jelly beans on the tray leads to the perception of a higher chance of picking one irrespective of the number of the other beans. One subject commented, "I picked the ones with the more red jelly beans because it looked like there were more ways to get a winner."

The absolute figures from the tamoxifen study showed that cases of invasive breast cancer were reduced from 261 per 10,000 without the treatment down to 133 per 10,000 with it. Ironically, had the ratio bias and mismatched framing been excluded in favor of the absolute numbers, the Breast Cancer Risk Tool's users would easily have been able to see that the total breast cancer cases prevented (128 per 10,000) hugely outweighed the uterine cancers caused by the treatment (14 per 10,000), with no manipulation of the original clinical data needed.

Regressive Attitudes

The majority of statistical misrepresentation in a medical context is likely done unconsciously by researchers who are unaware of some common statistical pitfalls. Clinical trials, for example, typically take a group of people who are unwell, give them a proposed treatment for their ailment, and monitor them for improvement to understand the effect of the medicine. If the symptoms are alleviated, it seems natural to give credit to the treatment.

Imagine, for example, recruiting a large number of people suffering from joint pain and asking them to sit still while you sting them with live bees. (Although this sounds absurd, it is a genuine alternative therapy known as apipuncture. Apipuncture has recently grown in popularity, in part due to its promotion by Gwyneth Paltrow on her Goop lifestyle website.) Now imagine that, miraculously, some of the sufferers' joint pain goes away—on average they start to feel better after the therapy. Can we conclude that apipuncture is an effective therapy for joint pain? Probably not. No scientific evidence supports apipuncture's efficacy for treating any disorder. Indeed, adverse reactions to bee-venom therapy are common and are known to have killed at least one patient. So how can we explain the positive results of our hypothetical trial? What causes the patients' improvement?

Conditions such as joint pain fluctuate in their severity over time. The sufferers recruited for the trial, especially for something as extreme and alternative as apipuncture, are likely at a particularly low ebb and are desperate for some resolution to their ailments when they sign up. If they receive treatment when their pain is at its worst, then at some time later they will highly likely be feeling better, irrespective of the benefits of the treatment. This phenomenon is known, ostentatiously, as "regression to the mean." It affects many trials in which the results have an element of randomness.

To understand better how regression to the mean works, consider the results of an exam. Take an extreme case in which students are asked to answer fifty yes/no questions on a subject they know nothing about. With the students effectively guessing entirely at random, the scores from the test could range from 0 all the way up to 50, but very few people will get only a small number of answers right, and very few will get almost none of the answers wrong. From the distribution of scores given in figure 20, clearly more people will get middling scores nearer to the average of 25. If we analyze the students in the top 10 percent, their scores will, by definition, be significantly higher than those of the whole-population average. Should we therefore expect these students to perform significantly above average when we retest them with fresh questions? Of course not. We would again expect their scores to be evenly distributed around a mean score of 25. The same would be true if we retested the bottom 10 percent. The individuals we picked out on the basis of their extreme scores in the first test will, on average, have reverted toward the mean in the second.

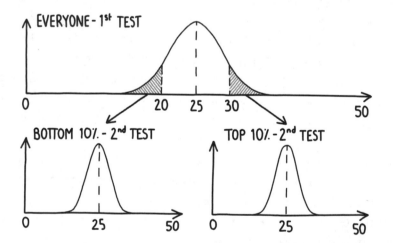

FIGURE 20: The spread of scores on a fifty-question "yes/no" exam. When those with the top 10 percent of scores (*right shaded region*) are retested, their mean score is the same as the mean overall score. The same is true for the bottom 10 percent (*left shaded region*). Both high-scoring and low-scoring populations have regressed toward the mean.

In real exams, skill and work ethic will play a significant role in determining a student's results, but an element of chance will probably also arise in the questions that come up and the subjects the students prioritized when studying for the exam. With a random component, regression to the mean will register its effect. The element of chance is especially prominent in multiple-choice exams, for which even a student without the requisite knowledge can guess at the right answer. In one study conducted in 1987, twenty-five test-anxious US students, who had performed unexpectedly poorly on the multiple-choice Scholastic Aptitude Test (SAT), were given the hypertension drug propranolol and retested. The *New York Times* reported the findings of the study as "a drug used to control high blood pressure has dramatically improved Scholastic Aptitude Test scores for students suffering unusually severe anxiety." Students taking propranolol improved their scores, on average, by a remarkable 130 points on a scale from 400 to 1,600. It appears, at first, that propranolol has a significant effect. However, even non-anxious students retaking the test improve their scores by around 40 points. When we consider that the students selected for the trial were chosen precisely because they performed worse than their IQ or other academic indicators had suggested they should, it wouldn't be surprising to see them significantly increase their scores even without propranolol, as a result of their regression to the mean.

In the absence of a similar set of poorly performing students who undertook the retest without taking the drug—the "control cohort"—it is impossible to determine the effects of the intervention. Based only on the treated cohort, it is tempting to attribute the improved performance to the effects of the drug. However, the results of the purely random multiple-choice test demonstrate that the regression of an extreme cohort toward the mean is a purely statistical phenomenon.

Avoiding the spurious inference of causality is of great importance in medical trials. One way to do this (as we have already seen in chapters 2 and 3) is to conduct a randomized controlled trial in which patients are allocated to one of two groups at random. As in the tamoxifen breast cancer trial, patients in the "treatment arm" receive the genuine treatment, and patients in the "control arm" receive a dummy or placebo treatment. If both the patients and the administrators of the treatment are kept in the dark about which arm of the trial the patient is in, the trial is known as double-blind—widely considered the gold standard for clinical trials. In a double-blind, randomized, controlled trial any difference between improvement in the control group and improvement in the treatment group can be attributed solely to the treatment, ruling out regression to the mean.

Historically, any improvement of patients in the control arm of the trial has been termed the placebo effect—the benefit of receiving what is perceived as a treatment, even if it's just a sugar pill. However, it is becoming increasingly clear that this effect is composed of two quite different phenomena. One, perhaps smaller, part is the genuine psychosomatic effect that makes patients feel better just because they believe they are being treated. This "true placebo" effect confers some genuine change in the patient's judgment of their symptoms. The psychosomatic benefit is larger if the patients know they are receiving the real treatment and, interestingly, even if the person administering the treatment knows, hence the reason for double-blinding.

The other, perhaps more significant, reason for the improvement of patients in the control arm is regression to the mean. This simple statistical effect confers no benefit to the patients at all. The only way to determine which is the more important of the two placebo components is to compare the effects of the dummy treatment with the effects of no treatment at all. These types of trial are often considered unethical, but enough studies have been done in the past to indicate that the majority of the so-called placebo effect is actually a result of regression to the mean—from which patients derive no benefit.

Many proponents of alternative medicine argue that, even if their treatment is nothing more than a placebo effect, the benefit of the placebo can be significant and is worth having. However, if the majority of the placebo effect is caused by regression to the mean, which provides no benefit to the patient, this argument doesn't hold water. Other alternative medicine gurus argue that, rather than putting stock in "artificial clinical trials," it is important to consider "real world" results—or, to paraphrase, "uncontrolled trial outcomes that focus solely on how patients' conditions change after treatment." Unsurprisingly, these "quacks" clutch at any argument that allows them to willfully misinterpret the effects of regression to the mean as a genuine causative benefit of their unscientific treatments. As Pulitzer Prize winner Upton Sinclair put it, "It is difficult to get a man to understand something, when his salary depends upon his not understanding it."

Away from medicine, regression to the mean also has far-reaching consequences for the interpretation of cause and effect in lawmaking. On October 16, 1991, thirty-two-year-old Suzanna Gratia sat down to eat with her parents in Luby's Cafeteria in Killeen, Texas. At its peak-time lunch hour the restaurant was unusually busy, with over 150 people packed around the square tables. At 12:39, George Hennard, an unemployed merchant seaman, accelerated his blue Ford Ranger pickup truck toward the restaurant and drove straight through the front window into the dining area. He immediately jumped out of the driver-side door and, holding his Glock 17 pistol in one hand and his Ruger P89 in the other, started to shoot.

Initially suspecting an armed robbery, Gratia and her parents dropped to the floor and upended the table as a makeshift barrier between them and the gunman. As shot after shot rang out, however, it became chillingly clear to Gratia that this man was not there to rob the restaurant; he was there to kill indiscriminately, and to kill as many people as possible.

The gunman approached within a few meters of Gratia's table, and she went for her purse. In it she kept concealed a .38-caliber Smith & Wesson she had been given for self-defense a number of years earlier. As she reached for the gun, however, her blood froze. She remembered that she had taken the cautious decision to leave the revolver under the passenger seat of her car so as not to fall foul of Texas's concealed-weapons law. She refers to it as "the stupidest decision of my life."

Gratia's father heroically decided that he would have to tackle the gunman before everyone in the restaurant was murdered. He sprang up from behind his table and rushed at Hennard. He didn't make it more than a few feet. Shot in the chest, he fell to the floor, mortally wounded. In his search for more victims, Hennard veered away from the table behind which Gratia and her mother were still concealed. Another customer, Tommy Vaughan, threw himself through a window at the back of the restaurant in a desperate bid to get away. Seeing the broken window as a potential escape route, Gratia grabbed her mother, Ursula, and insisted, "Come on, we gotta run, we gotta get outta here." Running as hard as she could, Gratia got herself quickly through the window and outside the restaurant unscathed. She turned back to ensure her mother had followed her, but found herself alone. Instead, Ursula had crawled to where her husband's body lay on the floor and cradled his head as he lay dying. Slowly, methodically, but surely enough, Hennard made his way back to where she sat and shot her in the head.

Gratia's parents were just two of the twenty-three that Hennard murdered that day. Twenty-seven others were wounded. At the time it was the worst mass shooting in US history. Gratia went on to give powerful testimony across the country in support of legalizing the bearing of concealed weapons. Prior to the Luby's massacre in 1991, ten states had "shall-issue" concealed-carry laws. These laws require that, provided an applicant satisfies a set of objective criteria, they must be issued a permit to carry a concealed weapon—with no discretion on the part of the issuer. Between 1991 and 1995, eleven more states passed similar laws, and on September 1, 1995, George W. Bush signed a law making Texas the twelfth.

Given that gun control is such a contentious issue in the United States, people had great interest in understanding the effects that these concealed-carry laws had on violent crime. Advocates of gun control suggested that more concealed weapons might lead to the fatal escalation of relatively minor disputes as well as an increase in the number of guns available to criminal factions. The gun-rights lobby suggested that the increased probability of a perpetrator's victim being armed might deter potential criminals, or at least allow citizens to attempt to end mass shootings sooner. The first studies comparing crime rates pre-introduction of the laws to those post-introduction seemed to indicate that rates of murder and violent crime were reduced in the immediate aftermath of the issuing of these concealed-carry laws.

However, two factors were typically neglected in these studies. The first was the decrease in violent crime across the whole country at the time when large numbers of concealed-carry laws were introduced. Between 1990 and 2001 increases in policing, rising numbers of incarcerations, and the receding crack cocaine epidemic all contributed to a fall in murders across the United States from a rate of around ten per hundred thousand per year to around six per hundred thousand per year. The prevalence of homicide decreased by almost exactly the same amount in states with and without concealed-carry laws. When murder rates in concealed-carry states are considered relative to the overall US murder rate, the suggested impact of concealed-carry laws diminishes significantly. Perhaps even more important is one study's finding that once regression to the mean was accounted for, the data "gives no support to the hypothesis that shall-issue laws have beneficial effects in reducing murder rates." States commonly passed concealed-carry laws in response to increasing levels of violent crime. That relative murder rates seemed to dip subsequent to the laws' introduction was seemingly unrelated to the concealed-carry laws. Instead, the laws were found to be associated with increased relative murder rates *prior* to their introduction. This gave a false impression of the laws' effectiveness as crime rates naturally dipped from their abnormally high levels.

Spotting the Spin

The debate on gun legislation in the United States still rages today. In the wake of the Las Vegas shooting, in October 2017, in which fifty-eight people were killed and hundreds more injured, Sebastian Gorka, recently relieved of his duties at the White House, participated in a roundtable debate on gun control. Gorka, who, as we saw at the start of the chapter, is no stranger to making bold, unsubstantiated claims, waded into a discussion about restricting the sales of firearms and their accessories and took the debate in an unexpected direction:

> It's not about inanimate objects. The biggest problem we have is not mass shootings, they are the anomaly. You do not make legislation out of outliers. Our big issue is black African gun crime against black Africans . . . black young men are murdering each other by the bushel.

Assuming Gorka was referring to African Americans, this sounds very much like a rehash of the bad statistics that were discredited earlier in this chapter. Gorka's repeated transgression highlights one of the situations in which you should be most on your guard against bad statistics: the serial offender. People who have shown a disregard for the accuracy of their figures once are unlikely to be more scrupulous in the future. The *Washington Post's* Glenn Kessler, one of the pioneers of political fact-checking, regularly analyzes and rates the statements of politicians on a scale from one to four "Pinocchios," depending on the degree to which they have bent the truth. The same names crop up over and over in his reports.

Other subtle signs can indicate a manipulated statistic. If presenters are confident of the veracity of their figures, then they won't be afraid to give the context and the source for others to check. As with Gorka's terrorism tweet, a contextual vacuum is a red flag when it comes to believability. Lack of details on survey results, including the sample size, the questions asked, and the source of the sample—as we saw in L'Oréal's advertising campaign—is another warning sign. Mismatched framing, percentages, indexes, and relative figures without the absolutes, as in the NCI's Breast Cancer Risk Tool, should set alarm bells ringing. The spurious inferences of a causative effect from uncontrolled studies or subsampled data—as often seen in the conclusions drawn from trials of alternative medicine—are yet more tricks to watch out for. If an initially extreme statistic suddenly rises or falls—as with gun crime in the United States—be on the lookout for regression to the mean.

More generally, when a statistic is pushed your way, ask yourself the questions "What's the comparison?" "What's the motivation?" and "Is this the whole story?" Finding the answers to these three questions should take you a long way toward determining the veracity of the figures. Not being able to find the answers tells its own story.

Mathematics provides multiple ways to be economical with the truth. The stats proclaimed in newspapers, championed in ads, or spouted by politicians are frequently misleading, occasionally disingenuous, but rarely completely incorrect. The seeds of the truth are usually contained within their figures, but rarely the whole fruit. Sometimes these distortions are a result of willful misrepresentation, while on other occasions the perpetrators are genuinely unaware of the bias they are imposing or the errors in their calculations. We will explore the catastrophic consequences of such genuine mathematical mistakes in more portentous contexts in the following chapter.

In his classic book, *How to Lie with Statistics*, Darrell Huff suggests that "despite its mathematical base, statistics is as much an art as it is a science." Ultimately, the degree to which we believe the stats we come across should depend on how complete a picture the artist paints for us. If it is a richly detailed, realist landscape with context, a trusted source, clear expositions, and chains of reasoning, then we should be confident in the veracity of the numbers. If, however, it is a dubiously inferred claim, supported by a minimalist single statistic on an otherwise empty canvas, we should think hard about whether we believe this "truth."

CHAPTER 5

WRONG PLACE, WRONG TIME

When Our Number Systems Let Us Down

Alex Rossetto and Luke Parkin were in their second year of sports science degrees at Northumbria University in the northeast of England. In March 2015, they signed up for a trial designed to investigate the effects of caffeine on exercise. The students were to be given 0.3 grams of caffeine, then put through their paces. Instead, because of a simple mathematical error, they found themselves in intensive care fighting for their lives.

After drinking the caffeine, dissolved in a mixture of orange juice and water, Rossetto and Parkin were to take part in a commonly administered exercise-performance trial, known as the Wingate test. The students were asked to get on an exercise bike and pedal as hard as they could to see how the caffeine affected their anaerobic power output. But shortly after ingesting the caffeine cocktail, before even getting near the bikes, the students started to feel dizzy, reporting blurred vision and heart palpitations. They were immediately rushed to the emergency room and put on dialysis machines. Over the coming days Rossetto and Parkin each lost nearly thirty pounds in weight.

The researchers administering the test had miscalculated the dose, stirring in an astonishing 30 grams of powdered caffeine instead of 0.3 grams. The students had ingested the equivalent of around three hundred cups of regular coffee in a few seconds. Ten grams has been known to be

THE MATH OF LIFE & DEATH

fatal in adults. Fortunately for Rossetto and Parkin, both were young and healthy enough to tolerate the massive overdose with few long-term effects.

The researchers administering the test had typed a decimal point into their mobile phones two spaces too far to the right, turning 0.30 grams into 30 grams. This is not the first time a misplaced decimal point has had dramatic effects. Other similar mistakes have had consequences ranging from funny to farcical, and even fatal.

In the spring of 2016, construction worker Michael Sergeant sent out an invoice for £446.60 after completing a week's work. A few days later he was surprised and excited to find £44,660 credited to his bank account after the director of the company Sergeant billed put the decimal point in the wrong place. For a few days Sergeant lived the life of a rock star. He spent thousands of pounds on a new car, drugs, drink, gambling, designer clothes, watches, and jewelry before the police caught up with him. Sergeant was forced to pay back the remaining money and to complete community service for his small-scale opportunism.

A more significant mistake occurred in Pinehurst, North Carolina, in 1985 when Sarah Simpson went to her dermatologist seeking treatment for vitiligo. The condition manifests as areas of unpigmented skin caused by the death of melanocytes, the cells in the skin that produce melanin. The doctor wrote a prescription for ".1 percent Oxsoralen lotion," which he suggested Simpson apply to the affected areas before returning to the office to be exposed to ultraviolet light. Oxsoralen is designed to make the skin more sensitive to the UV light and, as a result, encourages repigmentation in some patients. When Simpson went to pick up the Oxsoralen prescription from her local pharmacy, they were out of stock, but the pharmacist kindly agreed to procure some from a colleague at a nearby pharmacy. Simpson then applied the lotion to much of her body and went to the doctor's office for the UV light treatment. Two days later Simpson was admitted to the hospital with burns over 25 percent of her body.

The pharmacist had called his colleague and, misreading the doctor's handwriting on the prescription, incorrectly asked for "1 percent

Oxsoralen lotion" instead of the 0.1 percent lotion originally prescribed. Against best practices, the doctor had omitted the 0 preceding the decimal point, which made it significantly easier for the pharmacist to misread 1 percent instead of .1 percent. The highly concentrated drug made Simpson's skin so sensitive to the UV light that within minutes she had experienced UV exposure equivalent to several hours lying unprotected in the sun. Fortunately, although her skin was badly burned and blistered by the factor-of-ten error, Simpson's injuries did not prove fatal.

However, it was exactly that for eighty-five-year-old pensioner Mary Williams. In 2007, a community nurse visited Mrs. Williams as a favor to a colleague. The nurse was charged with delivering to Mrs. Williams, a diabetic, her insulin shot for the day. The nurse filled up her first insulin injecting pen with the required thirty-six "units" of insulin, but as she tried to inject it, the pen jammed. She tried again with the two other pens she had brought, but each one failed. Worried about what would happen to Mrs. Williams if she did not get her insulin, the nurse returned to her car to procure a regular syringe. Although the pens were marked simply in "units" of insulin and the syringe in milliliters, the nurse knew that each "unit" corresponded to 0.01 milliliters. She filled up the 1-milliliter syringe and injected it into Mrs. Williams's arm, then repeated the process three more times to complete the dosage. She did not question why she had to deliver multiple injections when a single dose had sufficed for her other patients. With the job finally completed, she left Mrs. Williams and continued on her rounds. Only later in the day did she realize her terrible mistake: instead of injecting 0.36 milliliters of insulin, she had given Mrs. Williams 3.6 milliliters—ten times too much. She immediately called a doctor, but by that time Mrs. Williams had already suffered a fatal, insulin-induced heart attack.

Although the mistaken protagonists in these stories can easily be lampooned for their obvious errors, the prevalence of such stories demonstrates that simple mistakes can and do happen, often with serious consequences. In part, the gravity of the repercussions of these mistakes is the fault of our decimal-place value system. In a number such as 222, each of the 2s represents a different number: 2, 20, and 200, with each being ten times bigger than the last. The scaling factor of ten makes placing a decimal point in the wrong place so serious. Perhaps, if we used the binary system—the system on which all our modern computerized technology is based, in which each place is only a factor of two larger than the last—we could avoid these errors. Injecting twice as much insulin or even prescribing four times as much caffeine might not have had such serious ramifications.

In this chapter we explore more of the costly errors that result from the systems that currently enumerate our everyday lives. We uncover the often-hidden influence of seemingly long-disused numerical systems that provide a window on our human history and shine a light on our biology. We discover the flaws that afflict them and look at the alternative systems being advocated that are helping to avoid common mistakes. We follow the natural selection of our counting systems down dead ends and along convergent paths that parallel the evolution of our human cultures. We unmask the mathematical thinking, paralleling the unconscious cultural biases that afflict us all, that is so deeply ingrained in our subconscious that we don't even recognize its restriction of our perspectives.

The Place

The current number system we adhere to is known as the decimal place value system. *Place value* because the same digit in a different *position* can represent a different numerical value. *Decimal* because the same digit in an adjacent position represents a number *ten* times bigger or smaller than its neighbor. The multiplication factor between places, ten, is known as the base. Why we use base ten rather than some other base is more of an accident of our biology than any well thought out plan. Although some of our forebears chose a different base, the vast majority of cultures that developed numerical systems (the Armenians, the Egyptians, the Greeks, the Romans, the Indians, and the Chinese, among others) chose decimal. The simple reason is that when we realized the need to enumerate, in much the same way as we teach our children today, we counted using our ten fingers.

Although base ten was the most common system adopted by our ancestors, some cultures chose other bases constructed from different aspects of our biology. The native Yuki people of California counted in base eight, using the spaces between their fingers as markers, rather than the fingers themselves. The Sumerians used base sixty, pointing to the twelve finger joints of the four fingers of the right hand, using the right thumb as a pointer, and keeping track of up to five sets of twelve (sixty) with the five fingers of the left hand. The Oksapmin people of Papua New Guinea use a system based on the number twenty-seven: starting with the thumb on one hand (1), traveling up and down the arm, taking in the nose (14), and finishing with the little finger on the other hand (27). So, while the ten fingers are by no means the only body parts to inspire a number system, they are the most obvious and hence the most commonly used by our ancestors when first developing mathematics.

Once a culture had established a system of counting, it opened up the possibility of developing higher mathematics that could be put to use for practical purposes. Indeed, many of the oldest human civilizations were well versed in sophisticated math. By the third millennium BCE, the Egyptians, for example, could add, subtract, multiply, and use simple fractions. Aptly, they knew the formula for the volume of a pyramid, and evidence suggests that they had stumbled across right-angled triangles with sides of length three, four, and five, a so-called Pythagorean triple, long before Pythagoras. The Egyptians used the common base ten, but had no place value system. Instead they had separate hieroglyphs for different powers of ten. These pictorial representations of numbers were not written in any particular order—the Egyptians knew how much each was worth by looking at the picture. The number 1 was just a single stroke, much as we have today, 10 was a cattle yoke, 100 was a coil of rope, 1,000 an ornate water lily, 10,000 a bent finger, 100,000 a tadpole, and 1,000,000 was the god Heh—the personification of infinity or eternity. A million was about as big as it got for the ancient Egyptians. If they wanted to represent the number 1999, they would draw one water lily, nine coiled ropes, nine yokes, and nine vertical dashes. Although awkward, the system works well enough for numbers under around a billion. However, if the

Egyptians had been able to fathom the number of stars in the universe (estimated at a huge 1,000,000,000,000,000,000,000,000,000 in our decimal place value system), they would have had to draw the god Heh a billion billion times — not really feasible.

The Romans had, in many ways, a far more advanced civilization than the Egyptians. Famously they popularized, on a large scale, inventions such as books, concrete, roads, indoor plumbing, and the concept of public health. However, their number system was more primitive. They used a system of seven symbols, I, V, X, L, C, D, and M to represent the numbers 1, 5, 10, 50, 100, 500, and 1,000 respectively. Appreciating that their numeral system was somewhat cumbersome, the Romans ensured that numbers were always written from left to right, largest to smallest, so the characters could then simply be added together. MMXV for example would represent 1,000 + 1,000 + 10 + 5 or 2015.

Because writing long numbers was so unwieldy, an exception to the rule was introduced. If a smaller number was ever found to the left of a larger number, it meant that it was to be subtracted from the larger number. The number 2019, for example, would be written MMXIX instead of MMXVIIII, the I subtracted from the final X to give 9, saving crucial characters. If this didn't complicate matters enough, the standardized rules and symbols of Roman numerals as we think of them today are probably not the same as those that the Romans actually used. The Etruscans, for example, may have used symbols such as I, Λ, X, ↑, and ✕, instead of I, V, X, L, and C, although even these are debated. The regulated symbols and rules for writing Roman numerals described above may have evolved over many centuries in post-Roman Europe. The systems used by real Romans were likely far less uniform.

Nevertheless, Roman numerals did not suffer the same annihilation as Egyptian hieroglyphs once the Roman Empire crumbled. Today, Roman numerals adorn many buildings to denote the year in which they were completed, allowing architects to lend an air of antiquity to a recently completed project. For that reason, the late 1800s were a particularly tough time for stone masons. The Boston Public Library bears the inscription MDCCCLXXXVIII — at thirteen characters, the longest Roman numeral of the last millennium—for the year 1888, in which it was completed. It's not just architects who feel that writing a number in Roman numerals gives it more gravitas. Fashion style guides propose that by using Roman numerals on your watch you are indicating you are more sophisticated than your average joe. Elizabeth II, the name of the longest-serving British monarch, does read less like a film sequel than Elizabeth 2. Films and television shows also make use of Roman numerals to denote the date of their production, but for different reasons. In the early days of film, since Roman numerals were harder to read quickly, the practice prevented most people from deducing too readily that they were watching recycled material, while still satisfying the filmmaker's copyright requirements.

Despite their niche longevity, Roman numerals never took over the world because their notational intricacies actively hindered the development of higher mathematics. Indeed, the Roman Empire is notable for its lack of eminent mathematicians and contributions to mathematics. As we have seen, each number in the Roman system is a potentially complex equation instructing the reader to add or subtract a series of symbols to come up with a result. This makes even the simple addition of two of these Roman numerals difficult. It wasn't possible, for example, to write down two numbers, one on top of the other, and add up the digits in each column, as we are all taught nowadays in our earliest math lessons. Two identical symbols in the same position in two different Roman numbers didn't necessarily mean the same thing. One can't simply subtract the digits of MMXV from the digits of MMXIX from right to left ($X - V$ gives 5, $I - X$ gives -9, etc.) to find that the gap between 2019 and 2015 is four years. Crucially, the Romans lacked the concept of a positional or place value number system.

Long before the Romans and the Egyptians, the people of Sumer, located in what is now modern Iraq, had a far more advanced number system. The Sumerians, often referred to as the originators of civilization, developed a wide range of technologies and tools for agricultural purposes, including irrigation, the plow, and possibly even the wheel. With the burgeoning of their agricultural society, it became necessary, for bureaucratic purposes, to measure plots of land accurately and to determine and record taxation. So around five thousand years ago the Sumerians invented the first place value system—a system whose most fundamental concepts would, eventually, spread across the globe. Numbers were written in a prescribed order. A symbol farther to the left represented a larger value than the same symbol placed farther to the right. In our modern place value system, in the number 2019, the 9 represents nine ones, the 1, one ten, the 0, no hundreds, and the 2, two thousands. Each time we move to the left, the same digit represents a number ten times larger. Although the Sumerians chose base sixty, they employed exactly the same principle. The rightmost column represented units, the next column to the left 60s, the next 3,600s, and so on. In the Sumerians' sexagesimal system, the digits 2019 would represent 9 ones, one 60, no 3,600s, and two 216,000s, giving a total, in decimal, of 432,069. Conversely, if the Sumerians had wanted to write the year 2019 in sexagesimal, it would have looked something like $\underline{33}$ $\underline{39}$, where the symbol $\underline{33}$ represents 33 lots of 60 (1980), and the symbol $\underline{39}$ represents the remaining 39 units.

The development of place value is arguably the most important scientific revelation of all time. It is no coincidence that the widespread European adoption of the base-ten Hindu-Arabic place value system (the system we still use today) in the fifteen century closely preceded the Scientific Revolution. Place value systems allow any number, no matter how large, to be tamed with just a few simple symbols. In the Egyptian and Roman systems, the position of a symbol had no global meaning. Instead the value was determined by the symbol itself, which meant that both cultures were hamstrung by the finite number of figures they could reasonably represent. The Sumerians, however, could represent any number they chose with their set of sixty symbols. Their sophisticated positional system allowed them to carry out advanced mathematics such as solving quadratic equations (which arise naturally in an agricultural context when apportioning land) and trigonometry.

Perhaps the primary reason the Sumerians used the sexagesimal system was because it significantly eased working with fractions and division.

Sixty has lots of factors: the numbers 1, 2, 3, 4, 5, 6, 10, 12, 15, 20, 30, and 60 all divide into 60 exactly with no remainders. Trying to divide a pound (made up of a hundred pence) or a dollar or a euro (made up of a hundred cents) between six people will cause a disagreement over who gets the four remaining pennies. A Sumerian mina, made of sixty shekels, could, however, be divided neatly between two, three, four, five, six, ten, twelve, fifteen, twenty, or even thirty people without causing a fight. Using the Sumerian's base sixty, it's also easy to measure and divide up a cake exactly and evenly between, for example, twelve people. One-twelfth, in the sexagesimal place value system, is just five-sixtieths. They would write this neatly as 0.5 (the first place after the point representing sixtieths in sexagesimal, instead of tenths in decimal), as opposed to the ugly 0.083333 . . . (8 hundredths, 3 thousandths, 3 ten-thousandths, etc.) in our decimal place value system. For that reason, just like a circular cake, Sumerian astronomers divided the arc of the night sky into 360 (that is, 6 × 60) degrees, helping them to make astronomical predictions.

The ancient Greeks, building on the Sumerian tradition, divided each degree into sixty minutes (denoted ') and each minute into sixty seconds (denoted "). Indeed, the word *minute* (think "mine-yute" not "min-it") just means an extremely small division (in this case of the circle), and *second* simply refers to second level of division of the degree. The sexagesimal place value system is still used in astronomy today and allows astronomers to capture the size of objects that are vastly different in the night sky. Because of its astronomical connections, the circular symbol for degrees, as in 360°, which is now also used for temperature, is thought originally to have symbolized the sun. Less romantically (and more mathematically), it's possible that it was natural to use a superscript $^{\circ}$ for degrees after ' and " had been used for its subdivisions minutes and seconds, completing the sequence O, I, II.

The Time

Although we may be less familiar with the minutes and seconds used in astronomy, a far better known sexagesimal system governs the rhythms of

THE MATH OF LIFE & DEATH

our everyday lives: time. From the moment we wake up to the moment we fall asleep, whether we know it or not, we are frequently thinking in sexagesimal. It is no coincidence that hours, the temporal divisions of our cyclic days, are also broken down into sixty minutes and each minute further into sixty seconds.

Hours themselves, however, are grouped into sets of twelve. Despite primarily using base ten, the ancient Egyptians divided the day into twenty-four segments: twelve day-hours and twelve night-hours, mimicking the number of months in the solar calendar. During daylight, time was recorded using sundials with ten divisions. Two twilight hours were added, one at either end of the day, for periods that were not yet dark, but for which the sundial was of no use. The night was similarly divided into twelve, based on the rising of particular stars in the night sky.

Because the Egyptians prescribed twelve hours in every daylight period, the length of their hours changed throughout the year as the amount of daylight changed with the seasons: longer in the summer and shorter in the winter. The ancient Greeks realized that to make significant progress with their astronomical calculations, equal segments of time would be a necessity, so they introduced the idea of dividing the day into twenty-four equal-length hours. However, not until the advent of the first mechanical clocks in fourteenth-century Europe did this idea really catch on. By the early 1800s, reliable mechanical clocks were widespread. Most cities in Europe divided their day into two sets of twelve equal hours.

The division of the day into two twelve-hour periods is still standard throughout much of the English-speaking world. Most countries, however, use the twenty-four-hour clock, which distinguishes, for example, eight in the morning (08:00) from eight in the evening (20:00) with numbers that are unambiguously twelve hours apart. The United States, Mexico, the UK, and much of the commonwealth (Australia, Canada, Egypt, India, etc.), however, still make use of the abbreviations a.m. (ante meridiem) and p.m. (post meridiem) or simply *before noon* and *after noon* to distinguish eight in the morning from eight in the evening. This discrepancy has occasionally been known to cause problems, especially to me.

When I was a graduate student, I was offered the opportunity to visit collaborators in Princeton. I am a bit of a nervous traveler, something I have inherited from my dad. Every time I set off from the house on an international trip, I hear him listing, "Money, tickets, passports," in an anxious voice in the back of my head. In much the same way, my recollection of the Pythagorean theorem for right triangles—"the square of the hypotenuse is equal to the sum of the squares of the other two sides"—still plays in the Irish accent of my high school math teacher, Mr. Reid.

Unsurprisingly, on my outbound trip from Heathrow I arrived an excessive four hours early for my flight. I bumped into my more relaxed and experienced supervisor, who was catching an earlier flight than mine, which was over two and a half hours later. My academic visit was productive, but my travel paranoia meant that I cut short my sightseeing trip to New York on my last day in the United States to make sure I got back to Princeton in enough time to get in a good night's sleep. That evening, with bags packed, room scoured, money, tickets, and passport checked and double-checked, I set my alarm for 4:00 a.m. to guarantee I wasn't late for my nine o'clock flight.

I duly woke up at four in the morning and boarded a train from Princeton. I arrived at Newark International Airport two and a half hours later. When I looked for my flight on the departure board, however, I couldn't find it. I scanned again and again, but the list skipped straight from the 8:59 to Saint Lucia to the 9:01 to Jacksonville. I went to the information desk and asked the lady sitting behind it about the flight. "I'm afraid the only flight we have going to London today leaves this evening, sir." I couldn't believe it. How had I made this mistake? I'd been so careful in my preparations, yet it seemed I'd overlooked that the flight I thought I was getting didn't even exist. Then it dawned on me. I asked the assistant what time the flight was leaving this evening. "Why, that would be the nine p.m. flight this evening, sir."

I had mixed up a.m. and p.m., a mistake that is simply not possible to make in the twenty-four-hour system. Fortunately, I had got it wrong in the right direction. My punishment was a fourteen-hour wait to board my flight, but the internet abounds with stories of people who have made this mistake in the opposite direction, completely missing their flight by twelve hours and having to fork out for a new ticket. This experience has done little to reduce my travel anxiety.

I found it difficult enough to arrive at the airport on time in the twenty-first century, but just imagine how challenging long-distance travel would have been with the confused and asynchronous time system of the early nineteenth century. By the 1820s, although most countries in Europe had broken their day into twenty-four equally sized hours, comparing the time between countries was so difficult as to be almost meaningless. Few nations had enforced a single time across the whole of their dominion, let alone coordinated with their neighbors. Bristol, in the west of the UK, could be as much as twenty minutes *behind* Paris, while London found itself six minutes *ahead* of Nantes, in the west of France. These discrepancies resulted, typically, because each city used a local time based on the position of the sun in the sky. Since Oxford is one and a quarter degrees farther west than London, the sun is at its peak there around five minutes later, setting local Oxford time five minutes behind London. The twenty-four hours corresponding to one 360° rotation of the Earth on its axis means that every degree of longitude is worth four minutes of time. Bristol, two and a half degrees west of London, was a further five minutes behind Oxford.

The problems that local time posed for long-distance travel on the burgeoning railway network led to the coordination of time across the UK. Using local time in the different cities of the UK led to disarray of timetables, and to several near misses due to confusion between drivers and signalmen. In 1840, the Great Western Railway adopted Greenwich Mean Time (GMT) across its network. The industrial cities of Liverpool and Manchester took up the cause in 1846. With the advent of telegraphy, times could be transmitted around the country almost instantaneously from the Royal Observatory in Greenwich, allowing cities to synchronize their clocks. Although the vast majority of the country quickly came on board with railway time, some cities, particularly those with strong religious traditions, refused to give up their "God-given" solar time in favor of the soulless pragmatism imposed by the railways. Not until 1880, when Britain's Parliament finally passed legislation, did the majority of solar-time stalwarts finally fall into line. Having said that, the bells of Tom Tower in Christ Church, a constituent college of the University of Oxford, still chime at five past the hour.

Italy, France, Ireland, and Germany all followed quickly in adopting a uniform time across their countries, with Paris time nine minutes ahead of GMT and Dublin twenty-five minutes behind. But the situation was not so simple in the United States. A single time across the fifty-eight degrees of longitude of the contiguous US mainland would not be practical for regions nearly four solar hours apart. In the winter, when the sun was setting in Maine, it would only just be lunchtime on the west coast. Clearly local time had some part to play, but in the middle of the 1800s the situation was extreme, with every major city holding to its own local time. Consequently, most of the railway companies operating across New England in 1850 had their own time, typically based on the location of their head offices or one of their more popular stations. At some busy junctions, up to five different times were respected. The confusion caused by this lack of uniformity is thought to have contributed to numerous accidents. After one particularly troubling incident in 1853, which lead to the deaths of fourteen passengers, plans were put in place to standardize time on the railroads of New England. Eventually, the whole of the United States was proposed to be divided into a series of time zones, each one hour behind the next, running east to west. On November 18, 1883, known to many across the country as the Day of Two Noons, station clocks were reset across the continent. The United States was divided into five time zones: Intercolonial, Eastern, Central, Mountain, and Pacific.

Inspired by the United States' subdivisions, in October 1884, at the International Meridian Conference in Washington, DC, Canadian Sir Sandford Fleming proposed to carve the whole Earth up into a series of twenty-four time zones, creating a globally standardized clock. The globe was divided by twenty-four imaginary lines, known as meridians, running from the South Pole to the North Pole. The universal day was to begin at midnight at the *prime* meridian in Greenwich. By the year 1900 almost everywhere on the Earth was part of some standard time zone, but not until 1986 did all countries measure their times with reference to the prime meridian, when Nepal finally set its clocks five hours and forty-five minutes ahead of Greenwich Mean Time. Having time zones that were offset from each other by regular portions of an hour saved a great deal of trouble and confusion, greatly simplifying timetabling and trade between neighboring countries. However, the introduction of time zones didn't completely eradicate confusion. Typically it meant that when mistakes were made, time calculations were not delayed by just a few minutes, but sometimes by up to an hour—a lag with the potential to cause disaster.

As the leader of the 26th of July Movement, in 1959, Fidel Castro, along with his brother Raúl and comrade Che Guevara, had overthrown the US-backed Cuban dictator Fulgencio Batista. Acting on his Marxist-Leninist philosophy, Castro quickly turned Cuba into a one-party state, nationalizing industries and businesses as part of sweeping social reforms. The US government could not countenance a Soviet-sympathizing communist state on its doorstep. By 1961, as the Cold War neared its zenith, the US hierarchy had come up with a plan to overthrow Castro. Fearing Soviet reprisal in Berlin, the American president, John F. Kennedy, insisted that the United States be seen to have no involvement in the coup. Consequently, a group of over a thousand Cuban dissidents, known as Brigade 2506, were trained for the invasion of Cuba in secret camps in Guatemala. The United States also stationed ten B-26 bombers (the type of plane with which the United States had armed Castro's predecessor) in nearby Nicaragua to assist with the invasion. On April 17, the exile brigade was to stage an invasion at the Bay of Pigs, on Cuba's southern coast. The idea was to spark an uprising with huge numbers of oppressed Cuban natives taking up the exiles' cause.

The plan was in trouble even before it could be enacted. On April 7, a full ten days before the anticipated attack, the *New York Times* got wind of the plans and ran a front-page story alleging that the United States had been training anti-Castro dissidents. Castro, now alert to the potential of invasion, took stringent precautions, imprisoning known dissidents who might assist in the uprising and readying his military. Nevertheless, on Saturday, April 15, two days before the invasion, the US B-26s flew to Cuba in an attempt to destroy Castro's air force. Their mission was an almost complete failure, destroying few of Castro's operational aircraft, and with at least one strafed B-26 ditched in the sea north of Cuba.

The botched mission sent Cuban foreign minister Raúl Roa storming to the United Nations. At an emergency session of the General Assembly, Roa claimed, correctly, that the United States had bombed Cuba. With the world's spotlight focused on the issue, Kennedy refused to risk providing further evidence of US involvement and canceled the air strike planned for the morning of April 17 to assist the exiles when they debarked.

Since Brigade 2506 was composed entirely of Cuban dissidents with no obvious link to the United States, Kennedy had plausible deniability with regard to their actions. On the morning of April 17, he sanctioned their landing on the beaches of the Bay of Pigs. They were confronted by twenty thousand well-prepared Cuban troops. Kennedy, again fearing international reprisals, refused to issue orders for Castro's army to be shelled or for planes to assist from above. By the evening of April 18, the exiles' invasion was on the ropes. In a last-ditch rescue attempt, Kennedy issued an order for the Nicaragua-based B-26s to strike at the Cuban military. The bombers would be protected by jets from a US aircraft carrier just over the horizon to the east of Cuba. The air strike was scheduled for 6:30 a.m. on the nineteenth.

As the allotted time approached, the jets set off to meet the B-26s, only to find that they had not arrived. The B-26s, working on Nicaraguan Central Time, arrived a whole hour later, at 7:30 Cuban Eastern Time. With no protection from the jets that had long since given up on their mission, two B-26s bearing American insignia were shot down by Castro's planes, proving beyond doubt US involvement in the attempted coup. The political ramifications of the simple time-zone mistake were huge, driving Cuba firmly into the arms of the Soviets and precipitating the Cuban Missile Crisis a year later.

By the Dozenal

The failure of the Bay of Pigs invasion is attributable, in part, to the division of the day, and hence the world, into two sets of twelve-hour time zones. However, the mistake would have been similarly disastrous had the Earth been broken up using a different base. With sixty or even just ten segments, Nicaragua's time zone would still have lagged behind Cuba's by roughly the same amount of time.

Many people think the base-twelve duodecimal or dozenal system is far superior to our predominant decimal system. Both the Dozenal Society of Great Britain and the Dozenal Society of America argue that the dozenal system's six factors, 1, 2, 3, 4, 6, and 12, in comparison to just four for decimal (1, 2, 5, and 10), give it an advantage — and I think they have a point.

My two children have taught me, through painful experience, that it's important to divide things equally. I'm sure they would prefer that they both had just one sweet each rather than one having five if the other had six. When we stopped at a service station on the way up to their grandparents', I bought a packet of Starburst. I passed the packet into the back for the kids to share. Little did I know that eleven Starburst were in the packet and that I had passed back an odd number of sweets for the kids to share between them. The fallout during the remainder of that long journey north is why I am careful now to buy only even numbers of sweets. Similarly, I have friends with three children who will only buy treats in triplicate. If you're the manufacturer of such child-focused products, you can maximize your audience and minimize the potential for upset by selling in sets of twelve, catering for families of one, two, three, four, six, or even twelve children. Similarly, next time you are dividing something up and it's important that everyone gets the same amount (cutting a cake at a children's party, for example), dividing into twelves will allow you increased flexibility in the number of people you can equitably accommodate. That said, if it isn't the sweets or the cake, I'm sure the kids will find something else to argue about.

The main basis for favoring dozenal over decimal is that, just as with the Sumerians' base sixty, more fractions have a "nice" closed representation in base twelve than in base ten. For example, in decimal, one-third has to be represented by the messy nonterminating decimal $0.33333\ldots$, whereas in dozenal one-third can simply be thought of as four-twelfths and written as 0.4 (the first place after the point representing twelfths in duodecimal). But why does this matter? Well, not having an exact representation of a number can make a difference when making repeated measurements. As an example, consider a meter-length of wood that you would like to divide into three equally long pieces to form the legs of a stool. Using a coarse decimal ruler, you approximate the first third as thirty-three centimeters and the second third again as thirty-three centimeters. But this leaves the final third at thirty-four centimeters. The resulting stool, with mismatched legs, might not be so comfortable to sit on. On a dozenal ruler, one-third, or equivalently four-twelfths, of a meter would be an exact marking, allowing you to divide the wood exactly into three equal-size legs.

Advocates of the dozenal system claim it would reduce the necessity for rounding off and hence mitigate a number of common problems. To some extent they're right. Although a wonky stool is perhaps of only minor inconvenience, the simple rounding errors that result from having to truncate the representation of numbers in our current decimal system can have more serious implications.

For example, a simple rounding error in a German election in 1992 nearly led to the denial of a seat in the parliament to the leader of the victorious Social Democratic Party when the Green Party's share of the vote was reported as 5.0 percent instead of 4.97 percent. In a completely different context, in 1982, a newly created Vancouver Stock Exchange index plummeted continuously over nearly two years, despite the market's bullish performance. Every time a transaction took place, the value of the index was rounded down to three decimal places, consistently knocking value off the index. With three thousand transactions per day the index lost around twenty points a month, undermining market confidence.

Imperial Rule

Despite its tendency to reduce errors associated with rounding, the upheaval and consternation it would cause means that the conversion to dozenal looks unlikely to be implemented by an industrialized nation anytime soon. However, many of the burgeoning industrialized nations of the past made extensive use of imperial measurement systems, which lean heavily on base twelve. A foot has twelve inches, with twelve lines to an inch. Originally, the imperial pound also had twelve ounces. The word *ounce* is derived from the same Latin word as *inch*, *uncia*, meaning a twelfth part. The imperial troy system, used for measuring precious metals and gemstones, still divides the troy pound into twelve troy ounces. The old British monetary pound comprised twenty shillings each made up of twelve pence. This meant that the 240-pence pound could be divided up evenly in twenty different ways.

Although the imperial system has some perceived advantages (the most commonly cited is forcing children to become familiar with obscure times tables), its non-uniformity (sixteen ounces to the pound, fourteen pounds to the stone, eleven cubits to the rod, four poppy seeds to the barleycorn, etc.) has largely been abandoned in favor of the decimal metric system. Today the United States remains in the company of Liberia and Myanmar as one of only three countries worldwide not to make extensive use of a metric system. Myanmar is currently trying to switch to metric. The United States' lack of conformity is based largely on skepticism and traditionalist stubbornness on the part of many of its citizens. In one episode of *The Simpsons*, so often a window into contemporary American life, Grampa Simpson rants, "The metric system is the tool of the devil. My car gets forty rods to the hogshead and that's the way I likes it."

The UK began its transition to metric in 1965 and is now a nominally metric country. Nevertheless, the UK has never quite relinquished the imperial measurements that it fathered. It still clings strongly to miles, feet, and inches for height and distance, pints for milk and beer, and stones, pounds, and ounces colloquially to measure weight. In February

2017, UK Secretary of State for Environment, Food and Rural Affairs and two-time Conservative-leadership candidate Andrea Leadsom even suggested that British manufacturers may be allowed to sell goods using the old imperial system after leaving the European Union. Although it appeals to a small minority of Grampa Simpsons overwhelmed by nostalgia for a bygone "golden era," switching back to imperial would leave the UK almost completely isolated in international trade. Much like the switch to dozenal, it would be incredibly expensive and time-consuming to implement and create mountains of needless bureaucracy. Bureaucracy and expense, combined with the reluctance to change of the people who live in the few remaining nonmetric countries, are also the primary reasons why the metric system has not yet been universally adopted. But while the United States remains the last industrial nation to use imperial units almost ubiquitously, it will continue to experience episodes in which it finds itself lost in translation.

On December 11, 1998, NASA launched its $125 million Mars Climate Orbiter, a robot designed to investigate the Martian climate and to be a communications relay to the Mars Polar Lander. In contrast to the Polar Lander, the Orbiter was never designed to reach the surface of Mars. Any approach closer than eighty-five kilometers would cause it to break up due to atmospheric buffeting. On September 15, 1999, after successfully negotiating its nine-month journey through the solar system, a final series of maneuvers was initiated to bring the Orbiter to the ideal altitude of around 140 kilometers above the surface of Mars. On the morning of September 23, the Orbiter fired its main thruster and then disappeared out of sight, forty-nine seconds earlier than expected, behind the Red Planet. It never came back into view. A post-accident investigation board concluded that the Orbiter had been on an incorrect trajectory that would have taken it to within fifty-seven kilometers of the surface, low enough for the atmosphere to destroy the fragile probe. Upon further investigation the board found that a piece of software, supplied by US aerospace and defense contractor Lockheed Martin, had been sending data about the

Orbiter's thrust in imperial units. NASA, one of the foremost scientific institutions of the world, was, unsurprisingly, expecting those measurements in standard international metric units. The error meant that the Orbiter fired its thrusters too vigorously and consequently became just another 338 kilograms (or, if you prefer, 745 pounds) of space junk as it fell apart deep in the Martian atmosphere.

Realizing that most of the rest of the world had converted to the metric system, and anticipating the sorts of mistakes that would afflict NASA, in 1970 Canada decided to make the move to metric. By the mid-1970s, products were labeled in metric units, temperature was reported in Celsius rather than Fahrenheit, and snowfall was measured in centimeters. By 1977 road signs had all been converted to metric, and speed limits were measured in kilometers per hour instead of miles per hour. For practical reasons, some industries took longer to convert to metric than others. In 1983, Air Canada's new Boeing 767s were the first to be calibrated in metric units. Fuel was measured in liters and kilograms rather than gallons and pounds.

On July 23, 1983, one of the newly revamped 767s landed in Montreal after a routine flight from Edmonton. Following a brief turnaround, which included refueling and a crew change, at 17:48, Flight 143 took off from Montreal on the return trip with sixty-one passengers and eight crew on board.

Cruising at an altitude of 41,000 feet or, as the electronic metric gauge read, 12,500 meters, Captain Robert Pearson set the plane to autopilot and relaxed. Around an hour into the flight, Pearson was startled by a loud beeping accompanied by flashing lights on the control panel. The warnings indicated low fuel pressure to the aircraft's left engine. Assuming a fuel pump had failed, an unperturbed Pearson switched off the alarm. Even without the pump, gravity should have continued to draw fuel into the engine. Seconds later the same alarm sounded, and warning lights again flashed on the dashboard. This time it was the right engine. Again Pearson turned off the alarm.

He realized, however, that with both engines potentially faulty, he would need to divert to nearby Winnipeg to get the plane checked out. As he had this thought, the left engine spluttered and failed. Pearson radioed Winnipeg, urgently informing them that he would need to undertake a single-engine emergency landing. As he desperately tried to restart the left engine, Pearson heard a noise from the control panel that neither he, nor his first officer, Maurice Quintal, had ever before heard. The second engine gave out, and the electronic flight instruments, powered by engine-generated electricity, went blank. The reason neither Pearson nor Quintal had ever before heard the alarm was because none of their training had dealt with the loss of both engines. The chances of both engines failing simultaneously was assumed to be negligible.

These engine failures were not even the first malfunctions the plane had experienced that day. When Pearson took charge of the plane earlier in the day, he had been informed that the fuel gauge was not working properly. Rather than grounding the flight and waiting twenty-four hours for the replacement part, Pearson decided that the amount of fuel required for the journey should be calculated by hand. Being a veteran pilot with over fifteen years' experience, this was nothing new to him. Based on average fuel efficiencies and leaving some margin for error, the ground crew worked out that the trip to Edmonton required 22,300 kilograms of fuel. Upon landing at Montreal a dipstick was used to find that the plane still contained 7,682 liters. This volume was multiplied by the fuel density, 1.77 kilograms per liter, to find that the plane already had 13,597 kilograms of fuel on board. This meant that the ground crew needed to add a further 8,703 kilograms or 4,917 liters. Perhaps Pearson should have noticed a problem at this point rather than later in the flight. When checking the ground crew's calculations, he might have remembered that the density of jet fuel is less than the density of water at one kilogram per liter, but then again Canada had only recently gone metric. Unfortunately, during Air Canada's protracted switch to metric, the figure 1.77 that the plane's documentation gave for the density of fuel was wrong: 1.77 converts liters of jet fuel to pounds, not kilograms. The correct figure should have been less than half that, 0.803, to convert liters to kilograms. Because of this error, Pearson really had only 6,169 kilograms of fuel already on board. This meant that the ground crew should have added 20,088 liters, four times more than the 4,917 liters they had calculated. Instead of the required 22,300 kilograms of fuel, Flight 143 took off with less than half that amount. The engines hadn't failed because of a mechanical fault. The 767 had simply run out of fuel.

The stricken plane continued to glide toward Winnipeg, with the only hope being that they could make a "dead stick," or unpowered, landing if their timing was just right. As luck would have it, Pearson was also an experienced glider pilot, so he set to calculating the plane's optimal glide speed to maximize their chances of making Winnipeg. However, as Flight 143 emerged from the clouds, the limited instruments available, powered by backup batteries, told Pearson that they would never make it. Pearson radioed air traffic control at Winnipeg with their situation. He was informed that the only airstrip that might be in range was at Gimli, approximately twelve miles from the plane's current position. In what seemed another stroke of luck, Quintal had been stationed at Gimli when he was a pilot in the Royal Canadian Air Force, so he knew the airfield well. What neither he, nor anyone in the control tower at Winnipeg, knew was that Gimli had since become a public airport and that part of the airport had been converted into a motorsports arena. At that very moment, the track was hosting a car race, with thousands of people in cars and camper vans watching from the surrounds of the runway.

As the plane approached the runway, Quintal attempted to lower the landing gear, but the hydraulic systems had given out when the engines stopped. Gravity was enough to pull the rear landing gear into place. Although the front landing gear also descended, it wouldn't lock in place: a piece of luck that would shortly be instrumental in saving many lives. With the engines silent, the race spectators had no idea that the free-gliding, hundred-ton tin can was approaching until it was almost upon them. As the plane hit the tarmac, Pearson braked as hard as he could, bursting two of the rear tires. Simultaneously the unlocked front landing gear collapsed back into the plane, unable to support its weight. The nose hit the ground, causing fountains of sparks to spray from the undercarriage.

The increased friction brought the plane to a swift halt, just a few hundred meters from the dumbfounded onlookers. Quick-thinking race stewards rushed onto the track to extinguish small friction-induced fires that had started in the nose, and all sixty-nine passengers and crew descended the emergency slides safely.

The Millennium Bug Bites

That Pearson managed to land the plane with barely any instruments or on-board computers was hugely impressive. As we move further into the twenty-first century, many modern technologies continue to experience the exponential acceleration of their development and propagation that we encountered in chapter 1. In particular, computers are progressively pervading our modern lives, and we are consequently becoming increasingly vulnerable to their failure. In the years preceding the turn of the new millennium, the so-called Millennium Bug loomed large over companies that relied on computer software for their operation. The software glitch was the legacy of an almost preposterously simple computer programming oversight of the 1970s and '80s.

If someone asks you for your date of birth, for brevity it's not unusual to give a six-digit answer. Some ambiguity may arise when a 10-year-old and a 110-year-old are asked to write down their birthdays, but the correct year for each can usually be inferred from the context. Computers, however, often operate without such context. In an attempt to be as economical as possible with memory (which was expensive in the early days of computing), most programmers employed a six-digit date format. Typically, they allowed their programs to assume the date belonged in the 1900s. This left scope for error if the date genuinely belonged in the next century. As the dawn of the new millennium approached, computer experts began to warn that many computer programs might be unable to distinguish between 2000 and 1900, or the first year of any other century for that matter.

When the clock eventually ticked over to midnight on January 1, 2000, little appeared to change. No planes fell out of the sky, no funds were wiped out, and no nuclear missiles were launched. The lack of dramatic and immediate consequences led to the widespread belief that fears of the effects of the Millennium Bug were blown way out of proportion. Some cynics even suggest that the computer industry may deliberately have exaggerated the scale of the problem to bring in a big payday. The opposing view is that the stringent preparation prior to the event helped to avert many potential disasters. There are many frivolous accounts of unremedied systems. Amusingly, the website of the US Naval Observatory, the organization responsible for keeping the nation's official time, showed the date "1 Jan 19100." However, some of the symptoms of the Millennium Bug were not so easy to laugh off.

In 1999, the pathology laboratory at the Northern General Hospital in Sheffield, England, was a regional hub for Down syndrome testing. Test results from pregnant women across the east of the UK were sent to Sheffield to be analyzed by their sophisticated computer model, run on the NHS computer system, PathLAN. The model took in a range of data about the women, including date of birth, weight, and the results of a blood test, to calculate a risk of their baby having Down syndrome. This assessment of risk helped the women decide how to proceed with their pregnancy, with high-risk mothers-to-be offered more definitive testing.

Throughout January 2000, staff in Sheffield found a number of isolated minor errors (relating to dates) in the PathLAN system, but these were quickly and easily rectified and not worried about. Later in the month, a midwife based at one of the hospitals served by the Northern General reported seeing far fewer high-risk Down syndrome cases than she would have expected. She reported the same findings three months later, but was on both occasions assured by staff in the lab that nothing was amiss. In May, a midwife from a different hospital reported similarly infrequent high-risk test results. Eventually, the manager of the pathology laboratory looked into the results. He quickly realized that something was amiss. The Millennium Bug had bitten with full force.

In the pathology lab's computer model, the mother's date of birth was used, with reference to the current date, to calculate her age. The mother's age is an important risk factor, with older mothers being significantly more likely to conceive a child with Down syndrome. After January 1, 2000, instead of a year of birth such as 1965 being subtracted from 2000 to give a mother's age of thirty-five, 65 was subtracted from 0, giving a negative age that the computer couldn't make sense of. Instead of triggering a warning, the nonsensical ages dramatically skewed the risk calculation, placing many older mothers in a lower-risk category than they should have been. As a consequence (in a similar misfortune to that which befell Flora Watson, the mother of baby Christopher, whose heart-wrenching "false negative" story we heard in chapter 2) over 150 women were sent letters incorrectly categorizing their unborn babies as being at low risk of having Down syndrome: false negatives. Of these, four women, who might otherwise have been offered further testing, gave birth to children with Down syndrome, and two other women had traumatic late abortions.

Binary Thinking

The computers on which we have increasingly come to rely work with the most primitive base possible — base two or binary. With decimal base ten, we require nine digits and a zero to represent any number. In the base-two system, we require only one digit other than zero. All binary numbers are strings of just ones and zeros. The word *binary* comes from the Latin *binarius*, meaning "consisting of two parts." In the binary place-value system, the same digit one place to the left of a neighbor represents a number larger by a factor of two, as opposed to a factor of ten, as we are used to in decimal. The first column on the right represents units, the second from the right 2s, the third 4s, the fourth 8s, and so on. To build a number such as eleven we need a one, a two and an eight, but no fours, so eleven has the binary representation 1011. There is an old mathematical joke: "There are only 10 types of people: those who understand binary and those who don't," 10, of course, representing the number two in binary.

Binary is the base of choice for computers, not because something is inherently nice about doing mathematics in binary, but because of the way computers are built. Every modern computer comprises billions of tiny electronic components called transistors, which communicate with one another, transferring and storing data. The voltage flow though a transistor is a good way to represent a numerical value. Rather than having ten reliably distinguishable voltage options for each transistor and working in decimal, it makes more sense to have just two voltage options: on and off. This "true or false" system means a small voltage can deliver a reliable signal that is not mistaken if it fluctuates slightly. By combining the true-or-false outputs of these transistors with logical operations such as "and," "or," and "not," mathematicians have shown that it is possible, in theory, to compute the answer to any mathematical calculation that has an answer, no matter how complex. Modern-day computers have gone a long way to realizing this theory practically. They are capable of performing incredibly complicated tasks by converting our requests into a series of ones and zeros and applying cold, hard logic to flip these bits back and forth until they provide a lucid answer. Despite the everyday miracles we achieve by enslaving the binary place-value system to the machines that live on our desks and in our pockets, at times this most primitive base has let its masters down.

Christine Lynn Mayes was just seventeen when she enrolled in the US army in 1986. She spent three years serving abroad in Germany as a cook before retiring from active duty, after which she returned home to study business at Indiana University of Pennsylvania, where she met her boyfriend David Fairbanks. In October of 1990, in need of money to support her studies, Mayes reenlisted in the army reservists. She joined the Fourteenth Quartermaster Detachment, a unit that would be charged with water purification. On Valentine's Day 1991, the unit was called to combat as part of Operation Desert Storm. Three days later, Mayes shipped out to the Middle East. On the day she left the United States, Fairbanks got down on one knee and proposed. Mayes willingly accepted his offer but, worried that she would lose it, declined to take the ring with her. "All right, then, it'll be here when you get back" were Fairbanks's last words to his fiancée before she left for Saudi Arabia. Fairbanks took the ring home with him and placed it atop a photo of

Christine next to his stereo. He would never have the chance to place the ring on her finger.

When the Fourteenth Quartermaster Detachment disembarked from the air base in the oil-rich city of Dhahran in Saudi Arabia, they were transported a short distance to their temporary barracks in the city of Al Khobar on the Gulf coast. The temporary building that housed Mayes's unit, as well as other American and British units, was little more than a corrugated-metal warehouse recently converted for human habitation. Six days after her arrival, on Sunday, February 24, Mayes rang home to tell her mother she had arrived safely and that her unit would soon be moving forty miles farther north toward the Kuwaiti border. The next day, having completed her shift, while others in her unit relaxed or worked out, Mayes slept, little suspecting that the events that would decide her fate had already been set in motion.

Despite the Iraqis' having launched over forty Scud missiles at Saudi Arabia during the Gulf War, fewer than ten had caused any significant damage. Most missiles that did reach Saudi Arabia were off course and landed in civilian areas, rather than on their intended military targets. In part, the Iraqis' lack of success was due to the Americans' Patriot missile system. The system was designed to detect incoming missiles and launch an "interception" to destroy them midflight. The system relied upon an initial radar detection, followed by a more detailed confirmatory detection, designed to ensure the target was a genuine missile rather than spurious noise detected by an overactive first radar. To make the more detailed detection, the secondary radar was sent the time and location of the first sighting together with an estimation of the projectile's velocity. These could then be used to produce a narrow window to search for the potential positions of the missile, allowing a more detailed verification.

For accuracy, the Patriot system counted time in tenths of a second. Unfortunately, although it has the nice short decimal representation 0.1, in binary one-tenth has an infinitely repeating expansion that looks something like 0.00011001100110011001100. . . . After the initial 0.0, the four digits 0011 just repeat over and over. No computer can store infinitely many numbers, so instead the Patriot missile approximated one-tenth using twenty-four binary digits. Because this number is a truncated representation, it is different from the true value of one-tenth by around one ten-millionth of a second. The programmers who wrote the code that governed the Patriot system assumed that such a small discrepancy would make no practical difference. However, when the system had been up and running for a long time, the error in the Patriot's internal clock accumulated to something significant. After about twelve days, the total error in the Patriot's recording time would be almost a second.

At 20:35 on February 25, the Patriot system had been running for over four days in a row. As Mayes slept, the Iraqi army launched a warhead atop a Scud missile toward the eastern coast of Saudi Arabia. Minutes later, as the missile crossed into Saudi Arabian airspace, the Patriot's first radar detected the missile and fed its data to the second radar for verification. When data was passed from one radar to the next, the time of detection was off by almost a third of a second. With the incoming Scud traveling at over sixteen hundred meters per second, its position was miscalculated by over five hundred meters. When the second radar searched the region in which it expected to find the missile, it drew a blank. The missile alert was assumed to be a false alarm and removed from the system.

At 20:40 the missile hit the barracks where Mayes was sleeping, killing her and twenty-seven of her colleagues and injuring almost a hundred others. This single attack, three days before the end of hostilities, was responsible for the deaths of one-third of all the US soldiers killed during the First Gulf War and could perhaps have been averted if computers spoke in a different language—with a different base.

No base, however, is capable of representing every number exactly with just a finite set of digits. With a different base, the Patriot missile detection error might have been avoided, but other errors would undoubtedly have occurred instead. So, despite the infrequent errors that it engenders, the energy and reliability advantages afforded by binary make it the most sensible base for our current computers. However, these advantages quickly evaporate if we try to employ binary in a societal context.

* * *

Picture yourself chatting to an attractive stranger you are pressed up against on a crowded bus. As your stop nears, you ask the person for his or her mobile phone number, and the person obligingly reels off some combination of ten digits of the form XXX-XXX-XXXX, the format common to all mobile phone numbers in the US. To achieve the same variety of numbers in binary, each mobile phone number would have to be at least thirty-three digits long. Imagine trying to take down 1110111 0011010110010011111111111111 before the bus reaches the stop and you have to get off. "Was that a one after the seventh zero or a zero?"

Of more immediate relevance is the potentially damaging binary thinking that pervades our society. From time immemorial, quick yes-or-no decisions have meant the difference between life and death. Our primitive brains had no time to calculate the probability that a falling rock would land on our heads. Coming face-to-face with a dangerous animal required a snap decision: fight or flight. More often than not a quick and overcautious binary decision was better than a slow, measured one that weighed up all the options. As we evolved into more complex societies, we retained these binary judgments. We fell back on stereotypes of our fellow human beings as good or bad, saints or sinners, friends or enemies. These classifications are crude, but they provided us with an easy shortcut that dictated how to react when faced with each individual. Over time, these stereotypes have become further entrenched by the binary caricatures that are a prerequisite in many popular dualist religions. Followers of these religions have no room to doubt what the characteristics of good and evil look like.

But for most of us nowadays, such quick decisions and absolute caricatures have little relevance. We have time to meditate more deeply on important life choices. People are too complex to be classified by a single binary descriptor, too ambiguous, too subtle. Binary thinking would leave no space on the page for some of our favorite characters: the Snapes, the Gatsbys or the Hamlets of the literary world. We like these mixed personas, entrenched in moral ambiguity, precisely because they reflect our own complex, flawed personalities. But still we reach for the comfortable certainty of binary labels to show the outside world who we are as people: we are red or blue, we are left or right, we are theists or atheists. We trick ourselves into self-defining as one of two options when so many more colors are in the spectrum.

In my own subject, mathematics, our biggest struggle is with such self-imposed false dichotomies: those who believe that they can do math and those who think they can't. There are far too many of the latter. But almost no one understands no math at all, almost no one cannot count. At the other extreme, for hundreds of years no mathematician has understood all of known mathematics. We all sit somewhere on this spectrum; how far we travel to the left or to the right depends on how much we think this knowledge can be useful to us.

Understanding the number systems around us, for example, gives us an insight into the history and culture of our species. These seemingly strange and often unfamiliar systems are not to be feared but to be celebrated. They tell us how our forebears thought and reflect aspects of their traditions. They also hold up a more tangible mirror to our most basic biology, demonstrating that mathematics is as intrinsic to us as the fingers on our hands or the toes on our feet. These systems teach us the language of our modern technology and help us to avoid simple mathematical mistakes. As we will see in the next chapter, by dissecting the mistakes we have made in the past, modern mathematics-based technology is (sometimes with dubious success) providing ways to avoid the same miscalculations in the future.

CHAPTER 6

RELENTLESS OPTIMIZATION

From Evolution to E-commerce,
Life Is an Algorithm

"In one hundred meters, turn right. . . . Turn right," instructed the disembodied voice from the satnav. With his wife and two of his children in the car, learner driver Roberto Farhat did just that. He had taken over the driving from his wife—a confident driver with fifteen years' experience—only minutes before. As he turned off the A6, a two-ton Audi traveling on the opposite lane careered into the passenger side of the car at forty-five miles per hour. In paying close attention to his satnav, Farhat had missed the road signs warning him *not* to turn right. Remarkably he walked away from the crash unhurt. His four-year-old daughter, Amelia, was not so lucky. She died in hospital three hours later.

We have all come to rely on devices such as satellite navigation systems to simplify our increasingly fraught lives. In determining the quickest route from A to B, satnavs have an intricate job to perform. On-demand calculation via an algorithm is the only feasible option to achieve this task. It would be challenging for one device to hold all possible routes between a distant pair of start and finish points. The vast number of beginnings and ends that might be requested magnifies the difficulty of the task astronomically. Given the difficulty of the problem, that satnav algorithms are rarely wrong is impressive. But when they do make mistakes, they can be disastrous.

An algorithm is a sequence of instructions that exactly specifies a job. The task could be anything from organizing your record collection to cooking a meal. The earliest recorded algorithms, though, were strictly mathematical in nature. The ancient Egyptians had a simple algorithm for multiplying two numbers together, and the Babylonians had rules for finding square roots. In the third century BCE, ancient Greek mathematician Eratosthenes invented his "sieve"—a simple algorithm for sifting out the primes from a range of numbers—and Archimedes had his "method of exhaustion" for finding the digits of pi.

In pre-Enlightenment Europe, increased skill in mechanical manipulation allowed for the physical manifestation of algorithms in tools such as clocks and, later, gear-based calculators. By the mid-nineteenth century this skill had progressed to such a degree that the polymath Charles Babbage was able to build the first mechanical computer, for which pioneering mathematician Ada Lovelace wrote the first computer programs. Indeed, Lovelace recognized that Babbage's invention had applications far beyond the purely mathematical calculations for which it had originally been designed: that entities such as musical notes or, perhaps most important, letters could be encoded and manipulated with the machine. First electromechanical, and then purely electrical, computers were harnessed for exactly this purpose by the Allies during World War Two to run algorithms that broke German ciphers. Although, in principle, the algorithms could have been implemented by hand, the prototype computers executed their commands with a speed and accuracy unmatchable by an army of humans.

The increasingly complex algorithms that computers now execute have become a vital part of the efficient handling of our day-to-day routines, from typing a query into a search engine or taking a photo on your phone, to playing a computer game or asking your digital personal assistant what the weather will be like this afternoon. We won't accept just any old solution either: we want the search engine to bring up the most relevant answer to our questions, not just the first one it finds; we want to know with accuracy the probability of rain at 5:00 p.m., so we can decide whether to take our coats with us to work; we want the satnav to guide us along the fastest route from A to B, not the first route it discovers.

Conspicuously absent from most definitions of an algorithm—a list of instructions to achieve a task—are the inputs and outputs, the data that give algorithms relevance. For example, in a recipe, the inputs are the ingredients, while the meal you serve up on the table is the output. For a satnav, the inputs are the start and finish points you specify, together with the map that the device holds in its memory. The output is the route the machine decides to take you on. Without these tethers to the real world, algorithms are just abstract sets of rules. As often as not, when the malfunction of an algorithm makes the news, incorrect inputs or unexpected outputs are the real story, not the rules themselves.

In this chapter we discover the math behind the unrelenting algorithmic optimization in our everyday lives, from the way our search results are ordered in Google to the stories pushed at us on Facebook. We expose the deceptively simple algorithms that solve difficult problems and upon which our modern tech giants rely: from Google Maps' navigation system to Amazon's delivery routes. We also step back from the computerized world of modern technology and deliver some algorithms directly into *your* control: the simple optimization algorithms you can use to get the best seat on the train or to choose the shortest queue at the supermarket.

Although some algorithms can perform tasks of unimaginable complexity, sometimes aspects of their performance are, to put it mildly, suboptimal. Tragically for the Farhat family, an out-of-date map caused their satnav to provide the wrong directions. The route-finding rules were not at fault, and had the map been up-to-date, it is likely the accident would never have happened. Their story illustrates the awesome power of modern algorithms. These incredible tools, which have pervaded and simplified many aspects of our daily lives, are not to be feared. But they must be treated with due reverence, and their inputs and outputs kept under close surveillance. With human supervision, however, comes the potential for censorship and bias. By considering what can happen when, for the sake of impartiality, manual control is curbed, we discover that prejudice may lie hidden, hard-coded within the algorithm itself: an imprint of the inclinations of its creator. No matter how useful algorithms can be, a little understanding of their inner workings, rather than blind faith in their error-free operation, can save time, money, and even lives.

The Million-Dollar Questions

In 2000, the Clay Mathematics Institute announced a list of seven Millennium Prize Problems, considered to be the most important unresolved problems in mathematics. The list includes the Hodge conjecture; the Poincaré conjecture; the Riemann hypothesis; the Yang-Mills existence and mass gap problem; the Navier–Stokes existence and smoothness problem; the Birch and Swinnerton-Dyer conjecture; and the P versus NP problem. Although their names mean little to many outside some relatively small subfields of mathematics, Landon Clay, the institute's eponymous major donor, indicated just how significant he believed each of these problems to be when he put up $1 million for the proof or disproof of each.

At the time of this writing, only the Poincaré conjecture has been solved. The Poincaré conjecture is a problem in the mathematical field of topology. You can think of topology as geometry (the math of shapes) with dough. In topology the actual shapes of objects themselves are not important; instead objects are grouped together by the number of holes they possess. For example, a topologist sees no difference between a football, a basketball, a baseball, or even a Frisbee. If they were all made of dough, they could theoretically be squashed, stretched, or otherwise manipulated to look like each other without making or closing any holes in the dough. However, to a topologist, these objects are fundamentally different from a rubber ring, a bike's inner tube, or a basketball hoop, which, bagel-like, each have a hole through the middle of them. A figure of 8 with two holes and a pretzel with three are different topological objects again.

In 1904, French mathematician Henri Poincaré (the same Poincaré who intervened to set right a mathematical travesty and exonerate Captain Alfred Dreyfus in chapter 3) suggested that the simplest possible shape in four dimensions was the four-dimensional version of a sphere. To explain what Poincaré meant by "simple," imagine making a loop of string around an object. If you can keep the string on the surface and pull it tight so that the loop disappears, then the article is, topologically, the same as a sphere. This idea is known as simple connectivity. If you can't always do this trick with the string, then you have a more complex topological object. Imagine threading the string, from underneath, through the center of a bagel and over the top. Now when you pull the string, the bagel gets in the way so that the loop never disappears. The bagel, with one hole, is fundamentally more complex than the football, with none. The result in three dimensions was already well-known, but Poincaré suggested that the same idea would hold in four dimensions of space. His conjecture was later generalized to state that the same idea should hold in any dimension. However, by the time the Millennium prizes were announced, the conjecture had been proved to be true in every other dimension, leaving only Poincaré's original four-dimensional conjecture unproved.

In 2002 and 2003, reclusive Russian mathematician Grigori Perelman shared three dense mathematical papers with the topology community. These papers purported to solve the problem in four dimensions. It took several groups of mathematicians three years to be sure of the veracity of his proof. In 2006, the year Perelman turned forty—the cut-off age for the prize—he was awarded the Fields Medal, mathematics' equivalent of the Nobel Prize. Although the awarding of the prize made a small ripple of news outside mathematics, it was nothing compared to the stories that started to circulate when Perelman became the first person ever to turn down the Fields Medal. In his rejection statement Perelman stated, "I'm not interested in money or fame. I don't want to be on display like an animal in a zoo." When the Clay Mathematics Institute were finally satisfied, in 2010, that he had done enough to merit the $1 million for solving one of their Millennium Prize Problems, he turned their money down too.

P versus NP

Although undoubtedly a hugely important piece of work in the field of pure mathematics, Perelman's proof of the Poincaré conjecture has

THE MATH OF LIFE & DEATH

few practical applications. The same is true for the majority of the other six Millennium Prize Problems that, at the time of this writing, remain unsolved. The proof or disproof of problem number seven, however—known succinctly, and somewhat cryptically, in the mathematical community as P versus NP—has the potential for wide-ranging implications, in areas as diverse as internet security and biotechnology.

It is often easier to verify a correct solution to a problem than it is to produce the solution in the first place. The P versus NP challenge, the most important of the open mathematical questions, asks whether every problem that can be checked efficiently by a computer can also be solved efficiently.

To draw an analogy, imagine you are putting together a jigsaw puzzle of a featureless image, such as a picture of clear blue sky. To try all the possible combinations of pieces to see if they fit together is difficult; to say it would take a long time is an understatement. However, once the jigsaw is complete, it is easy to check that it has been done correctly. More rigorous definitions of what efficiency means are expressed mathematically in terms of how quickly the algorithm works as the problem gets more complicated—when more pieces are added to the jigsaw. The set of problems that can be solved quickly (in what is known as Polynomial time) is called P. A bigger group of problems that can be checked quickly, but not necessarily solved quickly, is known as NP (which stands for Nondeterministic Polynomial time). P problems are a subset of NP problems, since by solving the problem quickly, we have automatically verified the solution we find.

Now imagine building an algorithm to complete a *generic* jigsaw puzzle. If the algorithm is in P, then the time taken to solve it might depend on the number of pieces, its square, its cube, or even high powers of the number of pieces. For example, if the algorithm depends on the square of the number of pieces, then it might take four (2^2) seconds to complete a two-piece jigsaw, one hundred (10^2) seconds for a ten-piece jigsaw, and ten thousand (100^2) seconds for a hundred-piece jigsaw. This sounds like a relatively long time, but it's still in the realm of only a few hours. However, if the algorithm is in NP, then the time taken to solve it might grow exponentially with the number of pieces. A two-piece jigsaw might still take 4 (2^2) seconds to solve, but a ten-piece jigsaw could take 1,024 (2^{10}) seconds and a hundred-piece jigsaw 1,267,650,600,228,229, 401,496,703,205,376 (2^{100}) seconds—vastly outstripping the time elapsed since the Big Bang. Both algorithms take longer to complete with more pieces, but algorithms to solve generic NP problems quickly become unserviceable as the problem size increases. For all intents and purposes, P might as well stand for those problems that can be solved Practically and NP for Not Practically.

The P versus NP challenge asks whether all the problems in the NP class, which can be checked quickly but for which there is no known quick solution algorithm, are in fact also members of the P class. Could it be that the NP problems have a practical solution algorithm, but we just haven't found it yet? In mathematical shorthand, does P equal NP? If it does, then, as we shall see, the potential implications, even for everyday tasks, are huge.

Rob Fleming, the protagonist of Nick Hornby's classic nineties novel, *High Fidelity*, is the music-obsessed owner of the secondhand-record store Championship Vinyl. Periodically, Rob reorganizes his enormous record collection according to different classifications: alphabetical, chronological, and even autobiographical (telling his life story through the order in which he bought his records). Quite apart from being a cathartic exercise for music lovers, sorting enables data to be interrogated quickly and reordered to display its different nuances. When you click the button that allows you to toggle between ordering your emails by date, sender, or subject, your email service is implementing an efficient sorting algorithm. eBay implements a sorting algorithm when you choose to look at the items matching your search term by "best match," "lowest price," or "ending soonest." Once Google has decided how well web pages match the search terms you entered, the pages need to be quickly sorted and presented to you in the right order. Efficient algorithms that achieve this goal are highly sought after.

One way to sort a number of items might be to make lists with the records in every possible permutation, then check each list to see if the order is correct. Imagine we have an incredibly small record collection comprising one album each by Led Zeppelin, Queen, Coldplay, Oasis, and Abba. With just these five albums we already have 120 possible orderings. With six we have 720, and with ten there are already over 3 million permutations. The number of different orderings grows so rapidly with the number of records that any self-respecting record fan's collection would easily prohibit them from being able to consider all the possible lists: it's just not feasible.

Fortunately, as you probably know from experience, sorting out your record collection, books, or DVDs is a P problem — one of those for which there is a practical solution. The simplest such algorithm is known as bubble sort and works as follows. We abbreviate the artists in our meager record collection to L, Q, C, O, and A and decide to organize them alphabetically. Bubble sort looks along the shelf from left to right and swaps any neighboring pairs of records that it finds out of order. It keeps passing over the records until no pairs are out of order, meaning that the whole list is sorted. On the first pass, L stays where it is as it comes before Q in the alphabet, but when Q and C are compared, they are found to be in the wrong order and swapped. The bubble sort carries on by swapping Q with O and then with A as it completes its first pass, leaving the list as L, C, O, A, Q. By the end of this pass, Q has been "bubbled" to its rightful place at the end of the list. On the second pass, C is swapped with L and A is promoted over O, so that O now resides in the correct place: C, L, A, O, Q. We need two more passes before A makes its way to the front and the list is alphabetically ordered.

With five records to sort we had to trawl though the unsorted list four times, doing four comparisons each time. With ten records we would have to undertake nine passes, with nine comparisons each time. This means that the amount of work we have to do during the sort grows almost like the square of the number of objects we are sorting. If you have a large collection, then this is still a lot of work, but thirty records would take hundreds of comparisons instead of the trillions upon trillions of possible permutations we might have to check with a brute-force algorithm listing all the possible orders. Despite this huge improvement, the bubble sort is typically derided as inefficient by computer scientists. In practical applications, such as Facebook's news feed or Instagram's photo feed, where billions of posts have to be sorted and displayed according to the tech giants' latest priorities, simple bubble sorts are eschewed in favor of more recent and more efficient cousins. The "merge sort," for example, divides the posts into small groups, which it then sorts quickly and merges together in the right order.

In the buildup to the 2008 US presidential election, shortly after declaring his candidacy, John McCain was invited to speak at Google to discuss his policies. Eric Schmidt, Google's then CEO, joked with McCain that running for the presidency was much like interviewing at Google. Schmidt then asked McCain a genuine Google interview question: "How do you determine good ways of sorting one million thirty-two-bit integers in two megabytes of RAM?" McCain looked flummoxed, and Schmidt, having had his fun, quickly moved on to his next serious question. Six months later, when Barack Obama was in the hot seat at Google, Schmidt threw him the same question. Obama looked at the audience, wiped his eye, and started off, "Well, er . . ." Sensing Obama's awkwardness, Schmidt tried to step in, only for Obama to finish, looking Schmidt straight in the eye, "No, no, no, no, I think the bubble sort would be the wrong way to go," to huge applause and cheers from the assembled computer scientists in the crowd. Obama's unexpectedly erudite response—sharing an in-joke about the inefficiency of a sorting algorithm—was characteristic of the seemingly effortless charisma (afforded by meticulous preparation) that characterized his whole campaign, eventually bubbling him all the way up to the White House.

With efficient sorting algorithms at hand, it's great to know that next time you want to reorder your books or reshuffle your DVD collection it won't take you longer than the lifetime of the universe.

In contrast to this, some problems are simple to state, but might require an astronomical amount of time to solve. Imagine that you work for a big delivery company such as DHL or UPS and you have a number of packages to deliver during your shift, before dropping your van back at the depot. Since you get paid by the number of packages you deliver and not the time you spend doing the deliveries, you want to find the quickest route to visit all of your drop-off points. This is the essence of an old and important mathematical conundrum known as the traveling salesman problem. As the number of locations to which you have to deliver increases, the problem gets extremely tough extremely quickly in a "combinatorial explosion." The rate at which possible solutions increase as you add new locations grows faster even than exponential growth. If you start with thirty places to deliver to, then you have thirty choices for your first drop-off, twenty-nine for your second, twenty-eight for your third, and so on. This gives a total of $30 \times 29 \times 28 \times \ldots \times 3 \times 2$ different routes to check. The number of routes with just thirty destinations is roughly 265 nonillion—that's 265 followed by thirty zeros. This time though, unlike with the sorting problem, there is no shortcut—no practical algorithm that runs in polynomial time to find the answer. Verifying a correct solution is just as hard as finding one in the first place, since all the other possible solutions need to be checked as well.

Back at delivery headquarters, a logistics manager might be attempting to assign the deliveries to be made each day to a number of different drivers, while also planning their optimal routes. This related task, the vehicle-routing problem, is even harder than the traveling salesman one. These two challenges appear everywhere—from planning bus routes across a city, collecting the mail from postboxes, and picking items off warehouse shelves, to drilling holes in circuit boards, making microchips, and wiring up computers.

The only redeeming feature of all of these problems is that, for certain tasks, we can recognize good solutions when they are put in front of us. If we ask for a delivery route that is shorter than one thousand miles, then we can easily check if a given solution fits the bill, even if it is not easy to find such a path in the first place. This is known as the decision version of the traveling salesman problem, for which we have a yes-or-no answer. It is one of the NP class of problems for which finding solutions is hard, but checking them is easy.

We can find exact solutions, despite difficulties, for some specific sets of destinations, even if it is not possible more generally. Bill Cook, a professor of combinatorics and optimization at the University of Waterloo in Ontario, spent nearly 250 years of computer time on a parallel super-computer calculating the shortest route between all the pubs in the United Kingdom. The giant pub crawl takes in 49,687 establishments and is just forty thousand miles long—on average one pub every 0.8 miles. Long before Cook started on his calculations, Bruce Masters from Bedford-shire in England was doing his own practical version of the problem. He holds the aptly titled *Guinness* world record for visiting the most pubs. By 2014, the sixty-nine-year-old had had a drink in 46,495 different watering holes. Since starting in 1960, Bruce estimates that he has traveled over a million miles in his quest to visit all the UK's pubs—over twenty-five times longer than Bill Cook's most efficient route. If you're planning a similar odyssey for yourself, or even just a local pub crawl, it's probably worth consulting Cook's algorithm first.

The vast majority of mathematicians believe that P and NP are fundamentally different classes of problems—that we will never have fast algorithms to dispatch salespeople or route vehicles. Perhaps that's a good thing. The yes-no "decision version" of the traveling salesman problem is the canonical example of a subgroup of problems known as NP-complete. A powerful theorem tells us that if we ever come up with a practical algorithm that solves one NP-complete problem, then we would be able to transmute this algorithm to solve any other NP problem, proving that P equals NP—that P and NP are in fact the same class of problems. Since almost all internet cryptography relies on the difficulty of solving certain NP problems, proving P equals NP could be disastrous for our online security.

On the plus side, though, we might be able to develop fast algorithms to solve all sorts of logistical problems. Factories could schedule tasks to operate at maximum efficiency, and delivery companies could find efficient routes to transport their packages, potentially bringing down the price of goods—even if we could no longer safely order them online! In the scientific realm, proving P equals NP might provide efficient methods for computer vision, genetic sequencing, and even the prediction of natural disasters.

Ironically, although science may be a big winner if P equals NP, scientists themselves may be the biggest losers. Some of the most astounding scientific discoveries have relied enormously on the creative thinking of highly trained and dedicated individuals, embedded deeply in their fields: Darwin's theory of evolution by natural selection, Andrew Wiles's proof of Fermat's Last Theorem, Einstein's theory of general relativity, Newton's equations of motion. If P equals NP, then computers would be able to find formal proofs of any mathematical theorem that is provable—many of the greatest intellectual achievements of humankind might be reproduced and superseded by the work of a robot. A lot of mathematicians would be out of a job. At its heart, it seems, P versus NP is the battle to discover whether human creativity can be automated.

Greedy Algorithms

Optimization problems, such as the traveling salesman problem, are so difficult because we are trying to find the best solution from an inconceivably large set of possibilities. Sometimes, though, we might be prepared to accept a quick, good solution rather than a slow, perfect one. Maybe I don't need to find the optimal way to minimize the space taken up by the things I put into my bag before I set off to work. Perhaps I just need to find a way that gets everything in. If this is the case, then we can start to take shortcuts in how we solve problems. We can use heuristic algorithms (commonsense approximations or rules of thumb) that are designed to get us close to the best solution for a wide range of variants of a problem.

One such family of solution techniques are known as greedy algorithms. These shortsighted procedures work by making the best local choice in an attempt to find globally optimal solutions. While they work quickly and efficiently, they're not guaranteed to produce the optimal solution or even a good one. Imagine you're visiting somewhere for the first time and you want to climb the highest hill around to see the lay of the land. A greedy algorithm to get to the top might first find the steepest incline at your current position, followed by taking a step in that direction. Repeating this procedure for every step will eventually lead you to a point where you face a decline in every direction. This means you've made it to the top of a hill, but not necessarily the highest hill. If you want to climb the highest peak to get the best view, then this greedy algorithm isn't guaranteed to get you there. It might be that the route to the top of the small hillock you've just scaled started off more steeply than the route that takes you up to the local mountain range, so you followed the hillock path erroneously, based on your heuristic myopia. Greedy algorithms can find solutions, but they're not always guaranteed to be the best solutions. For some problems, however, greedy algorithms are known to give the optimal solution.

A map held by a satnav can be thought of as a set of junctions that are connected by lengths of road. The problem that faces satnavs, to find the shortest route between two locations through a maze of roads and junctions, sounds as difficult as the traveling salesman problem. Indeed, the number of possible routes blows up astronomically quickly as the number of roads and junctions increases. Just a smattering of roads and a sprinkling of junctions is enough to push the number of possible routes into the trillions. If the only way to find the solution were to calculate all the possible routes and compare the total distance for each, then this would be an NP problem. Fortunately for everyone who uses a satnav, an efficient method—Dijkstra's algorithm—finds the solution to the "shortest path problem" in polynomial time.

For example, to find the shortest route from home to the cinema, Dijkstra's algorithm works backward from the cinema. If the shortest distance from home to all the junctions that are connected to the cinema by a single stretch of road is known, then the job becomes simple. We can simply calculate the shortest trip to the cinema by adding the lengths of the paths from home to the nearby junctions to the lengths of the roads connecting the junctions to the cinema. Of course, at the start of the process, the distances from home to the nearby junctions are not known. However, by applying the same idea again, we can find the shortest paths to these penultimate junctions by using the shortest paths from home to the junctions that connect to them. Applying this logic recursively, junction by junction, takes us all the way back to the house where we start our journey. Finding the shortest route through the road network simply requires us to repeatedly make good local choices—a greedy algorithm. To reconstruct the route we just keep track of the junctions we had to pass through to achieve this shortest distance. Some variation of Dijkstra's algorithm is likely crunching numbers under the hood when you ask Google Maps to find you the best route to the cinema.

When you arrive at the cinema and need to pay to park at the meter, the ticket machine probably won't provide change. If you have enough coins in your pocket, you probably want to make up the exact price as quickly as possible. In one greedy algorithm, which many of us will reach for intuitively, we insert coins sequentially, each time adding the largest-value coin that is less than the remaining total.

Most currencies, including those of Britain, Australia, New Zealand, South Africa, and Europe, share the 1-2-5 structure, with coins or notes increasing repeatedly in this pattern as the denominations of the currency get larger. The British system, for example, has one-, two-, and five-pence coins. Next come the ten-, twenty-, and fifty-pence coins, then £1 and £2 coins followed by a £5 note, and finally £10, £20, and £50 notes. So to make up fifty-eight pence in change in this system using the greedy algorithm, you would select the fifty-pence coin, leaving eight pence to make up. Twenty- and ten-pence would take you over the total, so next add a five-pence, followed by a two-pence, and finally a one-pence. For all the 1-2-5 currencies, as well as the US coinage system, the greedy algorithm described above does make up the total using the smallest number of coins.

The same algorithm isn't guaranteed to work in every currency. If, for some reason, there were a four-pence coin as well, then the last eight pence of the fifty-eight might have been made up more simply using two four-pence coins instead of a five, a two, and a one. Any currency for which each coin or note is at least twice as valuable as the next smallest denomination will satisfy the greedy property. This explains the prevalence of the 1-2-5 structure—the ratios of 2 or 2.5 between denominations guarantee that the greedy algorithm will work, while the simple decimal system is preserved. Because making change is such a common procedure, almost all the world's currencies have been converted so that they satisfy the greedy property. Tajikistan, with the five-, ten-, twenty-, twenty-five-, and fifty-diram coins, is the only country whose coinage doesn't satisfy the greedy property. It's quicker to make forty diram with two twenties than with the twenty-five-, ten-, and five-diram coins that the greedy algorithm would suggest.

On the subject of being greedy, have you ever tried asking for forty-three Chicken McNuggets at McDonald's? It may sound unlikely, but these battered, deep-fried poultry morsels have given rise to some interesting mathematics. In the UK, McNuggets were originally served in boxes of six, nine, or twenty. While having lunch with his son in McDonald's, mathematician Henri Picciotto wondered exactly which numbers of Chicken McNuggets he would not be able to order with combinations of the three boxes. His list contained 1, 2, 3, 4, 5, 7, 8, 10, 11, 13, 14, 16, 17, 19, 22, 23, 25, 28, 31, 34, 37, and 43. All the other numbers of McNuggets were achievable and known, from that day forth, as the McNugget numbers. The largest number you can't make with multiples of a given set of numbers is called the Frobenius number. So 43 was the Frobenius number for Chicken McNuggets. Sadly, when McDonald's started selling packs of four Chicken McNuggets, the Frobenius number tumbled to just 11. Ironically, even with the new box of four, the greedy algorithm fails when trying to make forty-three Chicken McNuggets (two boxes of twenty take you straight to forty, but there's no box of three), so even though it's now possible, asking for forty-three Chicken McNuggets at the drive-through might still prove to be a difficult problem.

Highly Evolved

When they work, greedy algorithms are highly efficient methods for solving problems. When they fail, though, they can be worse than useless. If you're keen to venture into the great outdoors to commune with nature by climbing the highest mountain around, getting stuck on top of a molehill in your back garden because you followed an inflexible greedy algorithm is less than optimal. Fortunately, a number of algorithms inspired by nature itself can help to get us over both the proverbial and literal hump.

One procedure, known as ant colony optimization, sends out armies of computer-generated ants to explore a virtual environment inspired by a real-world problem. When tackling the traveling salesman problem, for example, the ants walk between nearby destinations, reflecting real ants' ability to perceive only their local environment. If the ants find a short route around all the points, then they retrospectively lay pheromone on that route to guide other ants. The more popular, and correspondingly shorter, routes are reinforced and attract more ant traffic. As in the real world, the deposited pheromone evaporates, allowing the ants the flexibility to remodel the fastest route if the destinations change. Ant colony optimization is used to find efficient solutions to NP challenges such as the vehicle routing problem and to answer some of the toughest questions in biology, including understanding the way in which proteins fold from simple one-dimensional chains of amino acids into intricate three-dimensional structures.

Ant colony optimization is just one of a family of nature-inspired tools known as swarm intelligence algorithms. Despite communicating locally with only a small number of neighbors, flocks of starlings or schools of fish exhibit extremely rapid, but coherent, changes in direction. Information about a predator at one edge of a school of fish, for example, propagates quickly through to the other side of the group. By borrowing these local interaction rules, algorithm designers can dispatch huge flocks of well-connected artificial agents to explore an environment. Their rapid, swarm-like communication allows them to stay in touch with discoveries made by other individuals in their search for optimal surroundings.

Far and away nature's most famous algorithm is evolution. In its simplest form, evolution works by combining the traits of parents to produce children. The children that are better equipped to survive and reproduce in their environment pass on their characteristics to more offspring in the next generation. Sometimes mutations occur between generations, allowing new traits to be introduced, which may fare better or worse than those already in the population. Just three simple rules—select, combine, and mutate—are enough to generate the biodiversity that solves some of the planet's most difficult problems.

Before we get carried away by this eulogy to the panacea of biological evolution, it's important to recognize that evolutionary solutions are often good but rarely, if ever, perfect. On wildlife documentaries, or in articles about the natural world, we commonly hear of animals being "perfectly" adapted to their environment. From the desert-dwelling kangaroo rat that has evolved to go its whole life without ever drinking water, extracting all the moisture it needs from its food, to the notothenioid fish that have developed "antifreeze" proteins so they can survive in subzero oceans, evolution has produced animals that are brilliantly adapted to their challenging environments.

The search for perfection, however, should not be confused with evolution's blind exploration of the possibilities. Evolution typically finds a solution that works better than any previous solution for that environment, but it doesn't always come up with the best way to solve a problem.

The UK's red squirrel population provides a classic example. With its sharp claws, flexible hind feet, and long tail essential for balance, it is well adapted to climbing trees in the search for food. Its teeth grow continuously throughout its life, allowing a squirrel to crack open the hard outer shells of nuts with abandon and without its teeth ever wearing out. It was seemingly perfectly adapted to its environment, until an even better-suited relative arrived. The significantly larger gray squirrel finds and eats more food, as well as digesting and storing it more efficiently. Although the gray never fought or killed the red, its superior adaptation meant that it quickly dominated the broad-leafed woodlands of England and Wales, outcompeting the red and taking over its ecological niche. Our perception of many species' exemplary adaptation perhaps owes more to our limited imagination of what a genuinely "perfect" solution could look like, than it does to evolution finding a true optimum.

Despite that evolution does not necessarily find the best-possible solution, the central tenets of this best known of natural problem-solving algorithms have been plagiarized by computer scientists many times over, most notably in the "genetic" algorithms. These tools solve scheduling problems (including designing schedules for major sporting leagues) and provide good, if not perfect, solutions to difficult NP problems such as the "knapsack problem."

The knapsack problem imagines a trader who has a lot of goods to take to market in a rucksack with a fixed capacity. She can't take everything with her, so she has to make a choice. The different items each have different sizes and different profits associated with them. A good solution to the knapsack problem is a selection of goods that fit into the bag and give a high potential profit. Variations of the knapsack problem crop up when cutting shapes out of pastry or trying to be economical with your use of wrapping paper at Christmas. It appears when loading cargo ships and packing transport trucks. When download managers determine which data chunks to download and in what order to maximize the use of limited internet bandwidth, they are solving the knapsack problem.

A genetic algorithm starts by generating a given number of potential solutions to a problem. These solutions are the "parent" generation. For the knapsack problem, this parent generation comprises lists of packages that could fit in the knapsack. The algorithm ranks the solutions by how well they solve the problem. For the knapsack problem the ranking is based on the potential profit generated by that list of packages. Two of the best solutions—lists generating the most profit—are then *selected*. Some of the packages from one good knapsack solution are thrown out, and the remainder are *combined* with some of the packages from the other good solution. There is also the possibility of *mutation*—that a randomly chosen package might be removed from the knapsack and replaced with another. Once the first "child" solution in the new generation is produced, two more top-performing "parent" solutions are chosen and allowed to reproduce. That way the better solutions in the parent generation pass their characteristics on to more child solutions in the next generation. This combination process is repeated until enough children are produced to replace all the original solutions in the parent generation. Having had their turn, the parent solutions are then killed off, the new child solutions are promoted to parent status, and the whole selection, combination, mutation cycle starts again.

Because of the randomness in the way the child solutions are created, the algorithm doesn't guarantee that *all* the offspring it produces will be better than their parents. Many will be worse. However, by being selective about which of these children are allowed to reproduce—a virtual survival of the fittest—the algorithm dispenses with the dud solutions and allows only the best ones to pass on their characteristics to the next generation. As with other optimization algorithms, the solutions can hit a local maximum, for which any change will cause a decrease in fitness even though we have not yet reached the best-possible solution. Fortunately, the random processes of combination and mutation allow us to move away from these local peaks and to push on toward even better solutions.

The randomness that is such an important feature of genetic algorithms also has a role to play in our everyday lives. When you find yourself stuck in a rut, listening to the same songs by the same bands over and over, you might hit the shuffle button. In its purest form, shuffle will just pick a random song for you. It's like a genetic algorithm without the selection and combination stages, but with a high degree of mutation. It might be one way to find a new band you like, but you may have to wade through a heap of Justin Bieber or One Direction songs to get there.

Many music streaming services now provide far more algorithmically sophisticated ways to mix up your listening. If you've been playing a lot of the Beatles and Bob Dylan recently, a genetic algorithm might suggest you try a band that combines certain characteristics of the two—the Traveling Wilburys (the Bob Dylan–George Harrison supergroup) for example. By skipping songs or listening to them the whole way through, you signal a measure of their fitness, so the algorithm knows which "solutions" to work from in the future.

Netflix has plug-ins that select random films or box sets for you to watch, working from your previous preferences. Similarly, rafts of companies have recently sprung up offering to help relieve your food fatigue by sending you random selections of their products. From cheeses and wines to fruit and vegetables, you can start to optimize your gastronomic experience, exploring tastes you might not even have known were out there, while the caterers learn, based on your feedback, what to send you next. From fashion to fiction, companies are using tools from the evolutionary algorithms stable in an attempt to reinvigorate our everyday consumer experience.

Optimal Stopping

The mathematical underpinning of some of the optimization algorithms discussed above seems to suggest that they are solely the preserve of the tech giants who exploit them on such a huge scale for commercial gain. However, some more straightforward algorithms—albeit supported by sophisticated mathematics—can be employed to make small but significant improvements to our everyday lives. One such family is known as optimal stopping strategies and provides a way of choosing the best time to act to optimize the outcome of a decision-making process.

Pretend, for example, that you are looking for somewhere to take your partner out to dinner. You're both quite hungry, but you'd like to find somewhere nice. You don't want to dive into the first place you see. You consider yourself a good judge, so you'll be able to rank the quality of each restaurant relative to the others. You figure you'll have time to check out up to ten restaurants before your partner gets fed up with traipsing around. Because you don't want to look indecisive, you decide that you won't go back to a restaurant once you have rejected it.

The best strategy for this sort of problem is to look at and reject some restaurants out of hand to get a feel for what's out there. You could just choose the first restaurant you come to, but given you have absolutely no information about what's out there, the chances are only one in ten that you'll choose the best one at random. So you're better off waiting until you've judged a number of restaurants first, before choosing the first one you see that's better than all the others you've looked at so far. This restaurant-picking strategy is illustrated in figure 21. The first three restaurants are judged for quality, but then rejected. The seventh restaurant is better than all the others so far, so that's where you stop to eat. But is three the right number to reject? The optimal stopping question asks, "How many restaurants should you look at and reject just to get a sense of what's on offer?" If you don't look at enough, then you won't get a good feel for what's available, but if you rule out too many before taking the plunge, then your remaining choice is limited.

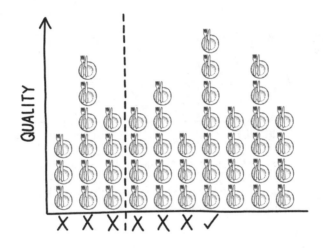

FIGURE 21: The strategy is to assess, but reject, every option up to a given cut-off point (dashed line) and then accept the next option you evaluate that is better than all the previous options.

The math behind the problem is complicated, but it turns out you should judge and reject roughly the first 37 percent of the restaurants (rounded down to three if there are only ten) before accepting the next one that is better than all the previous ones. More precisely, you should reject the fraction $1/e$ of the available options, where e is a mathematical shorthand for a number known as Euler's number. Euler's number is approximately 2.718, so the fraction $1/e$ is roughly 0.368 or, as a percentage, approximately 37 percent. Figure 22 illustrates how the probability of choosing the best of one hundred restaurants changes as you vary the number of restaurants you reject out of hand. Unsurprisingly, when you jump in and make a decision too soon, you are effectively guessing blindly, so the probability is low. Similarly, when you wait too long, it's likely you've already missed the best option. The probability of choosing the best option is maximized when you reject the first thirty-seven options.

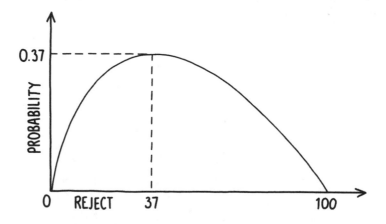

FIGURE 22: The probability of choosing the best option is maximized when we judge and reject 37 percent of the options before accepting the next one we assess that is better than all those we have previously seen. In this scenario, the probability you choose the best restaurant is 0.37 or 37 percent.

But what if the best restaurant was in the first 37 percent? In this case you miss out. The 37 percent rule doesn't work every time: it's a probabilistic rule. In fact, this algorithm is only guaranteed to work 37 percent of the time. That's the best you can do given the circumstances, but it's better than the 10 percent of times you would have chosen the best restaurant if you had just picked the first of ten at random, and way better than the 1 percent success rate if you had to choose at random between a hundred restaurants. The relative success rate improves the more options you have to choose from.

The optimal stopping rule doesn't work just for restaurants. The problem first came to the attention of mathematicians as the "hiring problem." If you have to interview a set number of candidates for a job one after the other, and at the end of each interview you have to tell the candidate whether he or she has got the job, then use the 37 percent rule. Interview

37 percent of the candidates and use those as your benchmark. Take the first interviewee after that who is better than all of the others you've seen so far and reject the rest.

When I reach the checkout at my local supermarket, I walk past the first 37 percent (four of the eleven), passively noting how long they are, then join the first queue thereafter that is shorter than all the others I've seen. If I am running, with a group of friends, for the packed last train after a night out and we want to find the carriage with the most spare seats so we can all sit together, we use the 37 percent rule. We head past the first three carriages on an eight-carriage train, remembering how empty they are, then take the first carriage after that that has more free seats than any of the first three.

Although rooted in realism, some of the above scenarios are a bit contrived, but they can be made more pragmatic. What happens if half of the restaurants you try don't have a table free? Then, understandably, you should spend less time rejecting restaurants out of hand. Instead of checking out the first 37 percent, look at just the first 25 percent before choosing the next one that's better than the ones you've encountered.

What if you decide you've got enough time to risk going back to an earlier carriage on the train, but the probability that it has filled up in the meantime is 50 percent? Because you're broadening your options by going back, you can afford to look for a bit longer—rejecting the first 61 percent of the carriages before choosing the next most empty carriage. Do make sure you get on the train before it starts to move off though.

Optimal stopping algorithms can tell you when to sell your house, or how far away from the cinema you should park to maximize your chances of getting a spot while minimizing the distance you have to walk. The caveat is that, as the situation becomes more realistic, the math becomes much more difficult and we lose the easy percentage rules.

Optimal stopping algorithms can even tell you how many people to date before you decide to settle down. You first need to decide how many partners you think you might get through by the time you'd like to settle down. Perhaps you might have one partner a year between your eighteenth and your thirty-fifth birthdays, making a total of seventeen potential partners to choose from. Optimal stopping suggests that you play the field for about six or seven years (roughly 37 percent of seventeen years) trying to gauge who's out there for you. After that you should stick with the first person who comes along who's better than all of the others you've dated so far.

Not many people are comfortable with letting a predefined set of rules dictate their love life. What if you find someone you're truly happy with in the first 37 percent? Can you coldheartedly reject the person because you're on an algorithmic love mission? What if you follow all the rules and the person you decide is best for you doesn't think that you're the best for him or her? What if your priorities change halfway through? Fortunately, in matters of the heart, as with other more obviously mathematical optimization problems, we don't always need to look for the very best solution, the single person who is the perfect fit—the One. Multiple people out there will likely be a good match and make us happy. Optimal stopping doesn't hold the answers to all of life's problems.

Despite the tremendous potential for algorithms to facilitate many aspects of our daily lives, they are far from the best solution to every challenge. Although an algorithm might simplify and accelerate a monotonous task, their use often has risks. Their tripartite nature—comprising inputs, rules, and outputs—means things can go wrong in three areas. Even if the users are confident that the rules are specified to their requirements, incautious inputs and unregulated outputs can lead to disastrous consequences, as online salesman Michael Fowler found out to his detriment. The American's algorithmically inspired retail master plan, which unraveled so abruptly in 2013, has its roots in Britain at the beginning of World War Two.

Keep Calm and Check Your Algorithm

Toward the end of July 1939 the dark clouds of war loomed ominously over Great Britain. In the air hung the possibility of heavy bombing, poison gas, or even a Nazi occupation. Worried about public morale, the British government resurrected a shadowy organization first instituted in the last year of World War One to influence news reporting at home and abroad: the Ministry for Information. Preempting an Orwellian amalgam of the Ministries for Truth and Peace, the new Ministry for Information would be responsible for propaganda and censorship during the war.

In August of 1939 the ministry designed three posters. Topped by a Tudor crown, the first read, "Freedom is in peril, defend it with all your might." The second read, "Your courage, your cheerfulness, your resolution, will bring us victory." By late August, hundreds of thousands of copies of these two posters had been printed, ready for use should war break out. They were distributed widely during the early months of the war to an audience, in the Great British public, who were largely left feeling either apathetic or patronized.

The third poster, which was printed at the same time, was held back for the severe and potentially demoralizing aerial bombardment that was expected. But by the time the Blitz actually began in September 1940, over a year after the start of the war, paper shortages, combined with the perceived condescension of the first two posters, led to the mass pulping of all three. The third poster was seen by almost no one outside the Ministry of Information.

In 2000, in the quiet market town of Alnwick, secondhand-book sellers Mary and Stuart Manley took delivery of a crate of used books they had recently purchased at auction. After emptying it out, they found a sheet of creased red paper at the bottom. As they unfolded it, they read the five words of the "lost" Ministry of Information poster: "Keep calm and carry on."

The Manleys liked the poster so much that they framed it and hung it on the wall of their shop, where it attracted the attention of their customers. By 2005 they were selling three thousand copies a week. But in 2008 the meme exploded into the global public consciousness. As recession gripped around the world, many sought to evoke the indomitable, stiff-upper-lipped demeanor of Brits who had previously struggled through tough times. "Keep calm and carry on" sold just the ticket. The message was transcribed to mugs, mouse mats, keychains, and every other piece of merchandise you can imagine. Even toilet paper did not escape the treatment. The message was transmuted in advertising campaigns for products as diverse as Indian restaurants ("Keep calm and curry on") and condoms ("Keep calm and carry one"). Almost any combination of "Keep calm and [insert verb] [insert noun]" seemed to resonate. Almost any.

This simple idea was harnessed by online merchandiser Michael Fowler. In 2010, Fowler's company, Solid Gold Bomb, was selling pre-printed T-shirts with around a thousand different designs when Fowler hit on an idea to increase the efficiency of his workflow. Instead of paying to store huge numbers of printed T-shirts, he would move to printing on demand. This would allow him to advertise many more designs, which would be printed only when an order was placed. Once the printing process was streamlined, he set to writing computer programs that would automatically generate the designs. Almost overnight, the number of Solid Gold Bomb's offerings jumped from a thousand to over 10 million. One such algorithm, created in 2012, took in a list of verbs and a list of nouns, combining them following the simple "Keep calm and [word from verb list] [word from noun list]" formula. The phrases generated by the procedure were then screened automatically for syntactical errors, superimposed on the image of a T-shirt, and listed for sale on Amazon for around $20 each. At Solid Gold Bomb's sales peak Fowler was selling four hundred T-shirts a day with phrases such as "Keep calm and kick ass" or "Keep calm and laugh a lot." But he had also automatically listed several T-shirts on the world's biggest online retailer with phrases such as "Keep calm and kick her" or "Keep calm and rape a lot."

Surprisingly, the phrases went almost unnoticed for a year. Then one day in March 2013, Fowler's Facebook page was suddenly inundated with death threats and allegations of misogyny. Despite his acting quickly to pull down the designs, the damage was already done. Amazon suspended Solid Gold Bomb's pages, sales dropped to nearly nothing, and despite staggering on for three months the company eventually went under. The algorithm Fowler designed, which seemed like such a good idea at the time, eventually cost him and his employees their livelihoods.

Amazon did not escape the episode unscathed either. The day after Solid Gold Bomb had issued their formal apology for the debacle, Amazon was still listing T-shirts with the slogans "Keep calm and grope a lot" and "Keep calm and knife her." A boycott of the retail giant was organized, with Lord Prescott, formerly the deputy prime minister of the United Kingdom, even joining in the twiticism: "First Amazon avoids paying UK tax. Now they're make [sic] money from domestic violence." Unsurprisingly, given the tech giant's heavy reliance on computer-automated procedures, this is just one of the many pitfalls of unsupervised algorithmic activity that the world's most valuable retailer has stumbled into.

In 2011, Amazon had already found itself the subject of algorithmic controversy as a result of automated pricing strategies. On April 8 that year, Michael Eisen, a computational biologist at Berkeley, asked one of his researchers to get the lab a new copy of *The Making of a Fly*, a classic, but out-of-print work in evolutionary developmental biology. When the researcher went to Amazon, he was pleased to find two new copies of the book for sale. When he looked more closely, though, he found that one of the books, sold by profnath, was on sale for $1,730,045.91. The other book, sold by bordeebook, was on sale for over $2 million. No matter how much he needed the book, Eisen couldn't justify the price, so instead he decided to watch the items to see if the price would come down. The next day when he checked the prices, he found that things were worse: both books were now priced at nearly $2.8 million. The day after they had risen to over $3.5 million.

Eisen quickly figured out a method to the madness. Each day profnath would set their price to be 0.9983 of bordeebook's offering. Later in the day bordeebook would scan profnath's listing and set its price to be roughly 1.27 times as large. Day after day bordeebook's price tag was inflated in proportion to its current size, causing it to grow exponentially. Profnath's listing lagged just behind. A human vendor controlling the prices would quickly have realized when the fees being asked for the books stretched beyond a sensible level. Unfortunately, this dynamic repricing wasn't being handled by a human, but by one of the range of repricing algorithms that are on offer to Amazon sellers. Apparently, no one thought to include a price-cap option on these algorithms, or if they did, the sellers decided against using it.

All the same, profnath's marginal underpricing strategy made some sense. It ensured their book was the cheapest available and consequently that it appeared at the top of the search list, while not giving away too much profit. But why would bordeebook choose an algorithm that continually priced their book out of the market so that it lay unordered, taking up space in their warehouse? It doesn't seem to make any sense, unless of course bordeebook never actually owned the book at all. Eisen suspects that bordeebook were trading on the trust and reliability indicated by their strong user ratings. If someone decided to buy the book from them, bordeebook would quickly buy the real copy from profnath and send it on to their buyer. The price hike allowed them to cover the postage they would have had to pay and still make a profit on the item.

Ten days after Eisen first spotted the exorbitant prices, they had spiraled further and reached $23 million. Sadly, on April 19, someone at profnath noticed the ridiculous price being asked for a twenty-year-old textbook and spoiled Eisen's fun by dropping the price back down to $106.23. The next day bordeebook's price was $134.97, roughly 1.27 times profnath's price, ready for the cycle to begin again. The price peaked again in August 2011, but this time at a mere $500,000, where it stayed unnoticed for the next three months. Apparently, someone learned their lesson and introduced a cap, although not a realistic one. At the time of writing you can find around forty listings for the book starting at the more reasonable price of around $7.

Despite the extortionate price tag, *The Making of a Fly* is not the most expensive item ever listed or sold on Amazon. In January 2010, engineer Brian Klug found a copy of a Windows 98 CD-ROM called *Cells* for sale on Amazon for nearly $3 billion (plus $3.99 postage and handling). The high price was presumably the result of another price spiral, with a second copy of the same CD-ROM listed by another seller, who had topped out at a comparatively modest $250,000. Klug entered his credit card details and purchased the item. A few days later Amazon emailed him to apologize that they couldn't fill his order. Disappointed, but probably equally relieved, Klug emailed back on the off chance that Amazon would honor the 1 percent credit against purchases made on the site using his Amazon credit card.

Flash Crash

Algorithmic price spirals, like those that affected Amazon, don't always coil upward. If you've ever invested in the stock market or even just put some savings in an account linked to it, then you will have heard the familiar refrain "The value of your investment can go down as well as up." Transactions on the stock market are increasingly being implemented as so-called algo-trading. Computers can sense and respond to changes in the market in a fraction of the time it would take their human counterparts. If a big order to sell a particular financial product flashes up on the screen, it may indicate that the price of that product is dropping and that traders are hoping to get rid of their assets at a good price before it drops further. In the time it takes humans to read the message and to click the button to sell their own assets, high-frequency algorithmic traders will already have sold theirs and the price will have dropped significantly. Human traders just can't compete. It is estimated that 70 percent of trading on Wall Street is now dealt with by these so-called black box machines. That's why the big-city traders and banks are increasingly turning to math and physics graduates, rather than brokers, to assist with writing and, perhaps more important, understanding these algorithmic traders.

On May 6, 2010, after an already poor morning for the markets, bit-part trader Navinder Sarao, operating from his London bedroom, turned on the bespoke algorithm he had recently finished modifying. The algorithm was designed to make a lot of money quickly by spoofing the market—having other traders believe and act on a trend in the market that wasn't actually there. His program was designed to rapidly place orders to sell a financial product known as E-mini futures contracts, but, before anyone could buy them, to cancel his order to sell.

Offering to sell his contracts at a price that was a little higher than the current best price ensured that no one, not even a fast-acting algorithm, would be tempted to take him up on his offer before his algorithm had a chance to cancel. When he executed the program, it worked like a charm. High-frequency algorithmic traders recognized huge numbers of sell orders coming in and decided to sell their own E-mini futures contracts before the price dropped — as it inevitably would if the market became saturated by so many sales. Once the price of the futures crashed, he proceeded to turn off his program and buy up the now-cheap contracts. Sensing the lack of sales, algorithmic traders rapidly regained confidence in and bought up the futures contracts, allowing the price to recover. Sarao made a killing.

His spoofing is estimated to have made him $40 million. His algorithm was hugely successful — perhaps too successful. High-frequency trading algorithms reacted to huge sales volumes in the futures market. In just fourteen seconds, the algorithms traded over 27,000 E-mini contracts, accounting for 50 percent of the day's total trading volume. They then began to sell other types of futures contracts to mitigate further losses. The fire sale then bled through into equities and out into the wider market. In the five minutes between 2:42 and 2:47 the Dow Jones dropped almost 700 points, bringing the total deficit for the day to nearly 1,000 points — the biggest single-day drop in the index's history — and wiping $1 trillion off the market. The high-frequency trading algorithms may not have caused the crash, but their unscrutinized, rapid trading certainly exacerbated it. Once the market bottomed out, however, and algorithmic confidence returned, they were also responsible for the rapid readjustment of most stocks back to near their opening values.

For nearly five years after the crash, US financial regulators blamed a whole raft of other factors for the flash crash. In 2015, however, Sarao was arrested and extradited to the United States for the part his scheme played in the 2010 crash. He pled guilty to illegally manipulating the market through spoofing and faces up to thirty years in prison, as well as having to pay back the money he earned through his illegal trading. Crime, it seems, even algorithmically assisted crime, doesn't pay.

Bang on Trend

Sarao's bedroom market manipulation illustrates just how straightforward it can be to employ algorithms for malign purposes. Too often we picture them simply as impartial sequences of instructions that can be followed dispassionately, forgetting that all algorithms are developed for a reason. Just because the rules themselves are predefined and can be executed impassively, it doesn't follow that the purpose for which they are being employed is unbiased, even if impartiality is the algorithm designer's original intention.

Twitter, often heralded as the bastion of transparency among social media platforms, uses a relatively straightforward algorithm to determine which topics are trending. The algorithm looks for sharp spikes in hashtag usage rather than promoting topics purely on the basis of high volume. This seems sensible: looking at the acceleration rather than just the rate of usage allows brief but important events, such as the request for blood donors (#dondusang—blood donation) or the offer of shelter for the night (#porteouverte—open door) in the aftermath of the coordinated 2015 terrorist attacks in Paris, to rise quickly to prominence. If high volume were the only trending criterion, then we would never hear of anything but Harry Styles (#harrystyles) and Game of Thrones (#GoT).

Unfortunately, this same set of rules means that social topics that build slowly are rarely catapulted to the prominence they might warrant. In September and October of 2011, throughout the Occupy movement, the hashtag #occupywallstreet never trended in the movement's native New York City despite its being Twitter's most popular hashtag during that period. Although they had less volume overall, more transient stories during that period, such as Steve Job's death (#ThankYouSteve) or Kim Kardashian's marriage (#KimKWedding), piqued attention in the right way to climb up Twitter's trending rankings. Remember that even genuinely pragmatic algorithms have hard-coded biases that influence the direction in which the spotlight is shone on the global stage.

Of perhaps greater concern are situations in which the results of seemingly independent algorithms are subject to human intervention. In May

2016, Facebook's "trending" news section was accused of anticonservative bias in an exposé article on the technology news website Gizmodo. Gizmodo heard testimony from a former Facebook news curator who claimed that right-wing stories on US political figures such as Mitt Romney and Rand Paul, among others, were being kept from Facebook's list of trending topics by human intervention. Even when conservative stories were organically trending on Facebook, they were allegedly not making it onto the trending list. In other cases, stories were purported to have been artificially "injected" into the trending list even if they weren't popular enough to merit inclusion.

In response to the accusations of political bias, Facebook decided to fire its trending editorial team and "make the product more automated." By divesting more power to the algorithm and removing a degree of human control, Facebook hoped to play on the perception of algorithmic objectivity. Just hours after their decision, the trending topics section was promoting a fake news story from the right wing, reporting that "closet liberal" Fox News anchor Megyn Kelly had been fired for her alleged support of Hillary Clinton. This would be just the first of a whole barrage of predominantly right-leaning fake news stories that would come to characterize Facebook's trending section over the next two years, making the allegations of anticonservative bias seem bland in comparison. The issue of reliability eventually led Facebook to pull the plug on the trending platform altogether in June 2018.

We place trust in supposedly impartial algorithms because we are wary of obvious human inconsistencies and inclinations. But although computers may implement algorithms in an objective manner following a predefined set of rules, the rules themselves are written by humans. These programmers might hard-code their biases, conscious or unconscious, directly into the algorithm itself, obfuscating their prejudices by translating them into computer code. The idea that we should be reassured about the neutrality of its trending news stories because Facebook, one of the world's foremost technology companies, had ceded power to one of its own algorithms doesn't hold water.

In common with Solid Gold Bomb's offensive T-shirts and Amazon's spiraling prices, Facebook's travails highlight the need for more, not less, human supervision. As algorithms become increasingly complicated, their outputs can become commensurately unpredictable and need to be policed with greater scrutiny. This scrutiny is not the sole responsibility of the tech giants, however. As optimization algorithms come to pervade more and more facets of our everyday lives, we—the frontline users of such shortcuts—need to shoulder part of the responsibility for ensuring the veracity of the outputs we are fed. Do we trust the source of the news stories we read? Does the route the satnav is suggesting make sense? Do we think the automated price we are being asked to pay represents value for the money? Although algorithms can provide us with information that facilitates vital decisions, in the end they are no substitute for our own subtle, biased, irrational, inscrutable, but ultimately human judgments.

When we investigate the tools at the forefront of the battle against infectious disease in the next chapter, we will find that exactly the same thesis holds true: although advances in modern medicine have gone a long way toward halting the spread of infectious diseases, mathematics shows that among the most effective ways we have to keep the lid on an epidemic are the simple actions and choices we make as individuals.

CHAPTER 7

SUSCEPTIBLE, INFECTIVE, REMOVED

How to Stop an Epidemic

During the Christmas holidays of 2014, the "Happiest Place on Earth" became a place of abject misery for many families. Hundreds of thousands of parents and children visited California's Disneyland over the holiday period, hoping to take home magical memories that would last a lifetime. Instead, some of them left with a memento that they hadn't bargained for: a highly infectious disease.

One of those visitors was four-month-old Mobius Loop. His mother, Ariel, and his father, Chris, were self-confessed Disneyland fanatics, to the extent that they got married there in 2013. As a trained nurse, Ariel was acutely aware of the dangers of exposing her prematurely born son's developing immune system to infectious disease. She confined her newborn almost exclusively to the house. She also insisted that anyone who wanted to visit Mobius before his first round of vaccinations, at two months, was up-to-date with their own jabs for seasonal flu, tetanus, diphtheria, and whooping cough.

In mid-January 2015, with his first round of vaccinations completed and their annual passes burning a hole in their pockets, Ariel and Chris decided to take Mobius to "experience the magic" at Disneyland. After a day watching parades and meeting supersized cartoon characters, the

Loops returned home, delighted by how much Mobius enjoyed his first adventure in the land of Disney.

Two weeks later, after a night spent struggling to get her son to sleep, Ariel noticed raised red marks on Mobius's chest and the back of his head. She took his temperature and found he was running a fever of 102 degrees Fahrenheit. Unable to control his temperature, Ariel rang her doctor, who instructed her to take the baby straight to the emergency room. When they arrived, they were met outside the hospital by an infection-control team wearing full protective clothing. Ariel and Chris were given their own masks and gowns and were rushed through a back entrance into a negative-pressure isolation room. Once inside, medical professionals inspected Mobius carefully, before asking Ariel to restrain him as they drew blood for a definitive test. Despite never having seen a case of the disease before, the emergency room staff all suspected the same thing: measles.

Due to the effectiveness of vaccination programs begun in the 1960s, few citizens of Western countries, including many medical professionals, have ever witnessed firsthand how severe the symptoms of measles can be. But travel to less developed countries such as Nigeria, where annual measles cases in the tens of thousands are routinely reported, and you might get a better picture of the disease. Complications can include pneumonia, encephalitis, blindness, and even death.

In the year 2000, measles was officially declared eliminated across the whole of the United States. Eliminated status meant that measles was no longer continuously circulating in the country and that any new cases were the result of outbreaks triggered by individuals returning from abroad. In the nine years from 2000 to 2008 there were just 557 confirmed cases of measles in the United States. But in 2014 alone there were 667 cases. As 2015 approached, the outbreak emanating from Disneyland, which affected the Loops and dozens of other families, spread rapidly across the country. By the time it died out, it had infected over 170 people in twenty-one states. The Disneyland flare-up is part of a trend of increasingly common large outbreaks. Measles is on the rise again in the United States and Europe, placing vulnerable people at risk.

* * *

Disease has afflicted humans ever since our hominin lineage first diverged from that of chimps and bonobos. Much of the story of our history has the often unwritten subplot of contagious disease running through it. Malaria and tuberculosis, for example, have recently been discovered to have affected significant swathes of the Ancient Egyptian population over five thousand years ago. From 541 to 542 CE, the global pandemic known as the Plague of Justinian is estimated to have killed 15 to 25 percent of the world's population of 200 million. Following Cortés's invasion of Mexico, the native population dropped from around 30 million in 1519 to just 3 million fifty years later; the Aztec physicians had no power to resist the previously unseen diseases brought over by the Western conquistadors. The list goes on.

Even today, in our increasingly medically advanced civilization, disease-causing pathogens are still sufficiently sophisticated that modern medicine has not been able to eliminate them from our daily lives. Most people experience the banality of the common cold almost every year. If you yourself have not had flu, then you surely know several people who have. Fewer in the developed world will have experienced cholera or tuberculosis, but these pandemic diseases are not uncommon in much of Africa and Asia. Interestingly though, even within communities in which disease prevalence is high, succumbing is not a certainty. Part of our morbid fascination with diseases is their seemingly random occurrence, visiting untold horrors on some while leaving others in the same community completely untouched.

A little-known, but highly successful, field of science is working in the background to unpick the mysteries of infectious disease. By suggesting preventive measures to halt the spread of HIV, and bringing the Ebola crisis to heel, mathematical epidemiology is playing a crucial role in the fight against large-scale infection. From highlighting the risks to which the growing anti-vaccination movement is exposing us, to fighting global pandemics, math is at the heart of the crucial life-and-death interventions that allow us to wipe diseases clean off the face of the Earth.

The Scourge of Smallpox

By the middle of the eighteenth century, smallpox was endemic through-out the world. In Europe alone it is estimated that four hundred thousand people a year died from the disease—up to 20 percent of all deaths on the continent. Half of those who survived were left blind and disfigured. Work-ing as a doctor in rural Gloucestershire, Edward Jenner had been witness to the truly held belief of his patients: that becoming a milkmaid could protect you from smallpox. Jenner deduced that the mild disease cowpox, which most milkmaids were exposed to, provided some immunity against smallpox.

To investigate his hypothesis, in 1796 Jenner carried out a pioneer-ing experiment into disease prevention that would be considered wildly unethical today. He extracted pus from a lesion on the arm of a milkmaid infected with cowpox and smeared it into a cut on the arm of an eight-year-old boy, James Phipps. The boy rapidly developed lesions and a fever, but within ten days was back on his feet, as fit and healthy as before the inoculation. As if having been infected by Jenner once was not enough, two months later Phipps submitted to Jenner's inoculating him again, this time with the more dangerous smallpox. After several days, when Phipps failed to develop symptoms of smallpox, Jenner concluded that Phipps was immune to the disease. Jenner named his protective process a *vaccination*, after the Latin word *vacca*, meaning "cow." In 1801, Jenner recorded his hopes for the discovery "that the annihilation of the Small Pox, the most dreadful scourge of the human species, must be the final result of this practice." Eventually, after a concerted vaccination effort by the World Health Organization nearly two hundred years later, in 1977, his dream became a reality.

The story of Jenner's development of vaccinations provides an indeli-ble link between smallpox and the history of modern disease prevention. Mathematical epidemiology also finds its roots in the attempt to diminish smallpox, but the subject's origins go back even further than Jenner.

*　　*　　*

Long before Jenner developed his idea of vaccination, in a desperate attempt to save themselves from the ever-increasing incidence of small-pox, the people of India and China practiced variolation. In contrast to vaccination, variolation involved exposing oneself to a small amount of material associated with the disease itself. In the case of smallpox, powdered scabs of previous victims were often blown up the nose or pus introduced into a cut in the arm. The aim was to induce a milder form of smallpox, which although still unpleasant was far less dangerous and would provide the patient with lifelong immunity from the severe symp-toms of the full-blown disease. The practice quickly spread to the Middle East and thence into Europe in the early 1700s, where smallpox was rife.

Despite its seeming effectiveness, variolation had its detractors. In some cases, the practice failed to protect patients from a second, more serious attack of smallpox as their immunity waned. Perhaps even more damaging to variolation's reputation were the 2 percent of cases in which patients died as a result of their treatment. The death of Octavius, the four-year-old son of the English monarch King George III, was one such high-profile case, which did little to improve the public's perception of the practice. Although a 2 percent mortality rate was still significantly lower than the 20 to 30 percent associated with the natural spread of the disease, critics argued that many variolated patients may never have been exposed to smallpox naturally and that widespread treatment was an unnecessary risk. It was also observed that variolated patients could also spread the full-blown disease just as effectively as naturally infected smallpox victims. In the absence of controlled medical trials, however, quantifying the effect of variolation and removing the lingering shadow over the procedure were not easily achieved.

This sort of public health issue piqued the interest of Swiss mathe-matician Daniel Bernoulli, one of the great unsung scientific heroes of the eighteenth century. Among his many mathematical achievements, Bernoulli's studies in fluid dynamics led him to propose equations that provide an explanation for how wings can create the lift required to allow planes to fly. Before he mastered advanced mathematics, however, Bernoulli's first degree was in medicine. His later studies into fluid flow combined with his medical knowledge led him to discover the first procedure that could be used to measure blood pressure. By puncturing the wall of a pipe with a hollow tube, Bernoulli could determine the pressure of the fluid running through the pipe by looking at how high it rose up the tube. In the uncomfortable practice that developed from his findings, a glass tube was inserted directly into a patient's artery. This method was not supplanted by a less invasive alternative for over 170 years. Bernoulli's broad academic background also led him to apply a mathematical approach to determine the overall efficacy of variolation, a question to which traditional medical practitioners could only guess the answer.

Bernoulli suggested an equation to describe the proportion of people of a given age who had never had smallpox and were hence still susceptible to the disease. He calibrated his equation with a life table, collated by Edmund Halley (of comet-spotting fame), which described the proportion of live births surviving to any given age. From this Bernoulli calculated the proportion of people who had had the disease and recovered, as well as the proportion who had succumbed. With a second equation, Bernoulli determined the number of lives that would be saved if variolation was practiced on everyone in the population. He concluded that with universal variolation, nearly 50 percent of infants born would survive to the age of twenty-five, which, although depressing by today's standards, was a marked improvement on the 43 percent if smallpox was allowed to rage freely in the population. Perhaps even more remarkably, he showed that this one simple medical intervention had the power to raise average life expectancy by over three years. For Bernoulli, the case for medical intervention by the state was clear. In concluding his paper he wrote, "I simply wish that, in a matter which so closely concerns the well-being of the human race, no decision shall be made without all the knowledge which a little analysis and calculation can provide."

Today, the purpose of mathematical epidemiology has not strayed far from Bernoulli's original aims. With basic mathematical models we can begin to forecast the progression of diseases and understand the effect of potential interventions on disease spread. With more complex models we can start to answer questions relating to the most efficient allocation of limited resources or tease out the unexpected consequences of some public health interventions.

The S-I-R Model

At the end of the nineteenth century, poor sanitation and crowded living environments in colonial India led to a series of deadly epidemics, including cholera, leprosy, and malaria, sweeping through the country and killing millions. The outbreak of a fourth disease, one whose very name has inspired terror for hundreds of years, would give rise to one of the most important developments in the history of epidemiology.

No one is entirely sure how the disease reached Bombay in August 1896, but the devastation it caused cannot be doubted. The most likely explanation seems to be that a merchant ship, harboring several highly undesirable stowaways, set sail from the British colony of Hong Kong. Two weeks later it docked at Port Trust in Bombay (now Mumbai). As the sweating longshoremen busied themselves unloading the ship's cargo in eighty-six-degree heat, several of the stowaways disembarked unnoticed and hurried off toward the city's slums. These free riders were harboring an unwanted cargo of their own, which would throw first Bombay, and then the rest of India, into chaos. The stowaways were rats carrying the fleas responsible for the spread of the bacterium *Yersinia pestis*: the plague.

The first cases of plague among Bombayites were detected in the Mandvi region encompassing the port. The disease spread unbridled through the city and, by the end of 1896, was killing eight thousand people a month. By the beginning of 1897, the plague had spread to nearby Poona (now Pune) and would subsequently spread across India. By May 1897, strict containment measures had seemingly caused the plague to die out. However, the disease would periodically return to haunt India for the next thirty years, killing over 12 million.

Into one such plague outbreak a young Scottish military physician, Anderson McKendrick, arrived in 1901. He would spend almost twenty years in India, carrying out research (remember, in chapter 1, we saw that McKendrick was the first scientist to show that bacteria increased to a carrying capacity according to the logistic growth model) and public health interventions and gaining a deeper understanding of zoonotic diseases—those diseases, such as swine flu, that can spread between animals and humans. Eventually his prowess in both research and practice would see McKendrick rise to be the head of the Pasteur Institute in Kasauli. Ironically, while in Kasauli he contracted brucellosis—a debilitating disease caused by drinking unpasteurized milk. As a result he was sent on several periods of medical leave back home to Scotland.

It was on one of these leave periods, inspired by an earlier meeting with fellow Indian Medical Service physician and Nobel Prize winner Sir Ronald Ross, that he decided to study mathematics. Mathematical study and research would dominate McKendrick's final years in India, before he was eventually invalided home permanently in 1920, after contracting a tropical bowel disease.

Back in Scotland, McKendrick became superintendent of the laboratory of the Royal College of Physicians of Edinburgh. Here he met the young and talented biochemist William Kermack. Not long after he met McKendrick, a devastating explosion left Kermack instantly and permanently blind. Despite this setback, his partnership with McKendrick flourished. Inspired by data on the plague outbreaks in Bombay, collected while McKendrick was in India, they conducted the single most influential study in the history of mathematical epidemiology.

Together they derived one of the earliest and most prominent mathematical models of disease spread. To make their model work, they split the population into three basic categories according to disease status. People who had not yet had the disease were labeled, somewhat ominously, as *susceptibles*. Everyone was assumed to be born susceptible and capable of being infected. Those who had contracted the disease and could pass it to susceptibles were the *infectives*. The third group was euphemistically referred to as the *removed* class. Typically, these people had had the disease and recovered with immunity or had died of it. The removed individuals no longer contributed to the spread of the disease.

This classic mathematical representation of disease spread is referred to as the S-I-R model.

In their paper, Kermack and McKendrick demonstrated the utility of their S-I-R model as it accurately re-created the rise and fall of the number of cases of plague in the 1905 outbreak in Bombay. In the ninety years since its inception, the S-I-R model and its variants have had great success in describing all sorts of other diseases. From dengue fever in Latin America to swine fever in the Netherlands and norovirus in Belgium, the S-I-R model can provide vital lessons for disease prevention.

Presenteeism, Predictions, and the Plague Problem

In recent years, the advent of zero-hour contracts and the increase in temporary employment—a hallmark of the burgeoning "gig" economy—have contributed to the rise in people coming to work while ill. While absenteeism has been extensively researched, the costs of "presenteeism" have only recently begun to be understood. Studies combining mathematical modeling and workplace attendance data have drawn some surprising conclusions. Measures implemented to reduce employee absence, including reducing paid sick leave, are causing a marked rise in people coming to work regardless of how bad they might be feeling, leading unintentionally to more illness and overall lowered rates of efficiency.

Presenteeism is particularly prevalent in health care and teaching. Ironically, nurses, doctors, and teachers feel so obligated to the large numbers of people they safeguard that they often put them at risk by coming in to work while under the weather. The hospitality industry, however, has perhaps the most acute presenteeism problem. One study found that in the United States alone over a thousand outbreaks of the vomiting bug, norovirus, were linked to contaminated food in the four years from 2009 to 2012. Over twenty-one thousand people fell sick as a result, and 70 percent of the outbreaks were linked to unwell food-service workers.

Five years after that study concluded, Chipotle Mexican Grill became a high-profile victim of the detrimental consequences of presenteeism. From 2013 to 2015 Chipotle was ranked as the strongest Mexican restaurant brand in the United States. Despite having paid sick leave, workers at many US Chipotle branches reported that managers required them to turn up to work when ill, under the threat of losing their jobs.

On July 14, 2017, Paul Cornell went out to enjoy a burrito in the Sterling, Virginia, branch of Chipotle. That same evening an unnamed food worker, despite stomach cramps and nausea, clocked in for work. Twenty-four hours later, Cornell was in the hospital, hooked up to an intravenous drip and suffering the extreme stomach pain, nausea, diarrhea, and vomiting consistent with a full-blown norovirus infection. One hundred and thirty-five other staff and customers also contracted the virus after visiting the restaurant. In the five days following the outbreak Chipotle's share price slumped, wiping over a billion dollars off the company's market value and leading its shareholders to file a class-action lawsuit against it. By the end of 2017, Chipotle didn't even make it into the top half of America's favorite Mexican restaurant chains.

The S-I-R model illustrates the importance of not coming in to work when unwell. By staying at home until fully recovered, you effectively take yourself from the infected class straight to the removed class. The model demonstrates that this simple action can reduce the size of an outbreak by diminishing the opportunities for the disease to pass to susceptible individuals. Not only that, but you also give yourself a better chance of a speedy recovery by not "working through the pain." The S-I-R model describes how, if everyone with an infectious disease followed this practice, we would all benefit through fewer preventable closures of restaurants, schools, and hospital wards.

Perhaps more than for its descriptive ability, however, the S-I-R model is vaunted for its predictive power. Instead of always looking at past epidemics, the S-I-R model allowed Kermack and McKendrick to look forward — to predict the explosive dynamics of disease outbreaks and understand the sometimes mysterious patterns of disease progression. They used their model to tackle some of the most hotly contested questions in epidemiology at the time. One such debate centered on the question "What causes a disease to die out?" Does a disease simply infect everyone in the population? Once the susceptible population is exhausted, perhaps the disease just has nowhere to go. Alternatively, perhaps the disease-causing pathogen becomes less potent over time, to the point at which it is unable to infect healthy individuals any more.

In their influential paper, the two Scottish scientists showed that neither of these is necessarily the case. Looking at their population at the end of a simulated outbreak, they found that some susceptible individuals always remained. This is, perhaps, in direct contrast to our intuition (fostered by movies and media scare stories), which would suggest that a disease dies out because no more people are left to infect. In reality, as infective people recover or die, contact between the remaining infectives and susceptibles becomes so infrequent that the infectives never get a chance to pass on the disease before being removed (recovering with immunity or dying). The S-I-R model predicts that, ultimately, outbreaks die out from a lack of infective people, not a lack of susceptibles.

In the small community of 1920s epidemic modelers, Kermack and McKendrick's S-I-R model was a towering contribution. It lifted the study of disease progression high above the rooftops of the purely descriptive studies that had gone before and allowed glimpses far off into the future. However, the windows of insight it provided were restricted by the narrow foundations on which the model was built: the numerous assumptions that limited the situations in which it could make useful predictions. These assumptions included a constant rate of human-to-human disease transmission; that infected people were also instantaneously infectious; and that population numbers didn't change. While useful for describing some diseases some of the time, these assumptions don't hold for the majority.

For example, ironically, the Bombay plague data that Kermack and McKendrick used to "validate" their model breaks many of these assumptions. For starters, the Bombay plague was not primarily passed from human to human, but spread by rats carrying fleas that were, in turn, carrying the plague bacterium. Their model also assumed a constant rate of transmission between infective carriers and their susceptible victims. In fact (as with the viral spread of the more trivial ice bucket challenge

from chapter 1), the plague in Bombay had a strong seasonal component, with flea density and bacterial abundance at dramatically higher
levels from January to March, leading to a consequent increase in the
transmission rate.

Nevertheless, future generations of mathematicians would adapt the
seminal S-I-R model, loosening its restrictive assumptions and expanding
the range of diseases to which mathematics could lend its insight.

One of the first adaptations made to the original S-I-R model was to
represent diseases that confer no immunity on their victims. One such
disease progression, typical of some sexually transmitted diseases, such as
gonorrhea, has no removed population at all. As soon as people recover
from gonorrhea, they can be infected again. Since no one dies from
gonorrhea, no one is ever "removed" by the disease. Such models are
typically labeled S-I-S, mimicking the progression pattern of an individual from susceptible to infective and back to susceptible again. Since
the population of susceptible people is never exhausted, but renewed
as people recover, the S-I-S model predicts that diseases can become
self-sustaining or *endemic*, even within an isolated population with no
births or deaths. In England, gonorrhea's being endemic has helped it
become the second most common sexually transmitted infection, with
over forty-four thousand reported cases in 2017.

Even more adaptations to the basic model are needed to properly
represent sexually transmitted diseases such as gonorrhea. Their progression pattern is not as simple as diseases such as the common cold
in which everyone can infect everyone else. With sexually transmitted
diseases, infectives typically only infect people corresponding to their
preferred sexual orientation. Since the majority of sexual encounters are
heterosexual, the most obvious mathematical model divides the population into males and females and allows infection only between these
two groups rather than among everyone. Models in which the bipartite
nature of heterosexual interactions is taken into account produce slower
disease spread than models in which it is assumed that everyone can

transmit the disease to everyone else irrespective of gender and sexual orientation. Such models of sexually transmitted diseases are, however, full of potential pitfalls.

HPV—More Than Just the Cancer Virus

The memories of my fifth birthday were still fresh when my mother was diagnosed with cervical cancer at age forty. She endured round after round of arduous and debilitating chemo- and radiotherapy. Thankfully, at the end of the grueling process, she was told that she was in complete remission. I was surprised to learn, later in life, that cervical cancer is one of the few cancers caused primarily by a virus—a cancer you can catch, typically through sexual intercourse. I find the idea that my father might have harbored the virus that caused my mother's cancer hard to take. He cared so devotedly for her when the cancer came back. Only his strength of will held our family together when she died a few weeks before her forty-fifth birthday. Even unknowingly, how could it have been him?

The vast majority of cases of cervical-cancer-causing human papillomavirus (HPV) infections *are* transmitted through sexual intercourse. Over 60 percent of all cervical cancers are caused by two strains of HPV. Indeed, HPV is the most frequent sexually transmitted disease in the world. Men can carry the virus asymptomatically and pass it to their sexual partners, contributing to cervical cancer's status as the fourth most common cancer in women, with around half a million new cases and quarter of a million deaths reported worldwide each year.

In 2006, the first revolutionary vaccines against HPV were approved by the US Food and Drug Administration. Unsurprisingly, given the high incidence of cervical cancer, hope surrounding the licensing of the vaccine was great. Studies undertaken in the UK around the time of the vaccine's deployment indicated that the most cost-effective strategy would be to immunize adolescent girls between the ages of twelve and thirteen, the likely future sufferers of cervical cancer. Related studies in other countries, considering mathematical models of the heterosexual transmission of the disease, confirmed that vaccinating females only was the best course of action.

However, these preliminary studies ultimately demonstrated that any mathematical model is only as good as the assumptions underpinning it and the data parameterizing it. The majority of these analyses neglected to include an important feature of HPV in their modeling assumptions: that the strains of HPV guarded against by the vaccine can also cause a range of noncervical diseases in both women and men.

If you've ever had a wart or a verruca, then you will have harbored at least one of five types of HPV. Eighty percent of people in the UK will be infected with one strain of HPV in their life. As well as causing cervical cancer, HPV types 16 and 18 contribute to 50 percent of penile, 80 percent of anal, 20 percent of mouth, and 30 percent of throat cancers. Famously, when in recovery from throat cancer, actor Michael Douglas was asked if he regretted his lifetime of smoking and drinking. He candidly told *Guardian* reporters that he had no regrets because his cancer had been caused by HPV that he contracted through oral sex. In both the United States and the UK, the majority of cancers caused by HPV are not cervical. Significantly, HPV types 6 and 11 also cause nine out of ten cases of anogenital warts. In the United States approximately 60 percent of the health-care costs associated with all noncervical HPV infections are for the treatment of these warts. Cervical cancer is an important part of the HPV narrative, but it is not the whole story.

In 2008, at the time the vaccine was first being rolled out, German virologist Harald zur Hausen was awarded the Nobel Prize in medicine "for his discovery of human papilloma viruses causing cervical cancer." The link to other cancers and diseases was somewhat ignored by the prize committee and most of the rest of the world. The one UK study that did account for noncervical cancers could not do so with any certainty because, at the time, the burden of the diseases and the impact of the vaccination against them was not properly understood. Most models suggested that by vaccinating a sufficiently high proportion of females, the prevalence of HPV-related diseases in unprotected males would also decline. The general public, perhaps aware only of HPV's role in cervical cancer—the common cancer that spreads like an infectious disease—accepted without question the decision only to vaccinate girls. Why should boys be vaccinated if they don't suffer from the headline HPV cancer?

But imagine the public outcry if a vaccination for the AIDS-causing human immunodeficiency virus (HIV) was developed and it was ordained that only women would be given the vaccination for free, in the hope that men would be protected through women's immunity. Quite aside from the issues associated with partial vaccination coverage and vaccine inefficiency, perhaps the first point that critics would make would be about the protection of gay men—should they be left with no defense against the deadly virus? The same argument holds true in the case of HPV. By neglecting homosexual relationships in their mathematical models, early studies ignored the effects of same-sex couplings. Models based on sexual networks including homosexual relationships have a higher rate of disease transmission than those that consider only heterosexual relationships. The prevalence of HPV in men who have sex with men is significantly higher than in the general population. In the United States, the incidence rate of anal cancer in this group is over fifteen times higher. At thirty-five per hundred thousand, it is comparable to the rates of cervical cancer in women *before* cervical screening was introduced and is significantly higher than current rates of cervical cancer in the United States. When models were recalibrated, taking into account homosexual relationships, new knowledge on the protection afforded against noncervical cancers, and up-to-date information on the length of protection that the vaccination provides, vaccinating boys as well as girls became a cost-effective option.

In April 2018, the UK's National Health Service eventually offered the HPV vaccination to homosexual men between the ages of fifteen and forty-five. In July of the same year, advice based on a new cost-effectiveness study recommended that all boys in the UK be given the HPV vaccination at the same age as girls. Thankfully, my daughter and my son will be afforded equal protection against catching and disseminating the virus that killed their grandmother. As this shows, conclusions drawn from even the most sophisticated of mathematical models are only as strong as their weakest assumptions.

The Next Pandemic

Another confounding factor with HPV infection is asymptomatic disease-carrying. People can harbor the virus, infecting others without showing symptoms themselves. For this reason, another adaptation that is commonly made to the basic S-I-R model, to represent diseases more realistically, is to include a class of people who, once infected, are capable of passing on the disease while remaining asymptomatic. The "carrier" class changes the S-I-R model into an S-C-I-R model and is vital for representing the transmission of many diseases, including some of the most deadly of our time.

Some patients experience a short period of flu-like symptoms a few weeks after contracting HIV. The severity of the symptoms varies broadly, and some carriers will not even notice anything wrong. Despite no outwardly obvious symptoms, the virus slowly damages the patients' immune system, leaving them open to opportunistic infections such as tuberculosis or cancers to which people with healthy immune systems might not succumb. Patients at the later stages of HIV infection are said to have acquired immune deficiency syndrome (AIDS). One of the main reasons why HIV/AIDS has become pandemic, meaning it has spread throughout the world and is still spreading, is this long incubation period. Carriers who are unaware that they have the virus spread the disease far more rapidly than people who know they are HIV positive. Each year, for the last thirty years or more, HIV has been one of the top causes of death by infectious disease worldwide.

HIV is thought to have emerged from nonhuman primates in Central Africa early in the twentieth century. Possibly as a result of humans handling infected primates caught for bushmeat, a mutated form of the simian immunodeficiency virus (SIV) jumped species into humans and was able to spread from human to human through the exchange of bodily fluids. Zoonotic diseases, which jump between species, like the original strains of HIV, present one of the biggest potential threats to public health.

In 2018, England's deputy chief medical officer, Professor Jonathan Van-Tam, singled out one such disease, the H7N9 virus—a new strain of bird flu—to be the most probable cause of the next global flu pandemic. The virus is currently highly prevalent in Chinese bird populations and has infected over fifteen hundred people. To put this into perspective, the Spanish flu, the most deadly pandemic of the twentieth century, infected roughly 500 million people worldwide. However, the mortality rate of the Spanish flu was only around 10 percent. H7N9 kills roughly 40 percent of those it infects. Fortunately, so far, H7N9 has not acquired the crucial ability to pass between humans, which would allow it to mushroom to the scale of the Spanish flu. Despite animal experiments suggesting that it is just three mutations away from being able to do so, perhaps, like its predecessor the bird-flu strain H5N1, it never will. The next global pandemic might not be from an emerging disease at all, but perhaps from one we have seen many times before.

Patient Zero

One afternoon in late 2013, two-year-old Emile Ouamouno was playing with some of the other children in the remote Guinean village of Melian-dou. One of the children's favorite haunts was a huge, hollow cola tree on the outskirts of the village—perfect for hiding in. The tree's deep and dark cavity also provided the ideal roost for a population of insect-eating free-tailed bats. While playing in the bat-infested tree, Emile came into contact either with fresh bat guano, or perhaps even face-to-face with a bat itself.

On December 2, Emile's mother noticed that her usually energetic young toddler was tired and lethargic. After feeling the heat of a fever emanating from his forehead, she took him to bed to recover. However, he soon started vomiting and excreting black diarrhea. He died four days later.

Having cared diligently for her son, Emile's mother also contracted the disease and died a week later. Emile's sister Philomène succumbed next, followed by their grandmother on the first day of the new year. The village midwife, who had cared for the family during their illnesses, unknowingly took the disease with her to neighboring villages and thence to the hospital in the nearest town, Guéckédou, where she sought treatment for her affliction. From there, one of many conduits for the spread of the disease was through a health worker who treated the midwife. She spread the virus to a hospital in Macenta, about fifty miles to the east, where she infected the doctor who attended her. In turn, he infected his brother in the city of Kissidougou, eighty miles to the northwest, and so the spread continued.

On March 18, 2014, the number of cases and their extent were becoming a serious worry. Health officials publicly announced the outbreak of an, as yet unidentified, hemorrhagic fever "which strikes like lightning." Two weeks later, upon identification of the disease, Médecins Sans Frontières called the scale of its spread "unprecedented." From this point on, Emile Ouamouno, an otherwise unremarkable child, would be transformed into someone the world would never forget. Tragically, he would become infamous as "Patient Zero": the victim of the first animal-to-human transmission of what had become the biggest, most uncontrolled Ebola outbreak of all time.

That we even know the progression of the disease is a tribute to the huge detail in which the epidemic was analyzed by scientists and health-care professionals placing themselves directly in its path. A method known as contact tracing allows epidemic experts to work their way backward through many generations of infected individuals, all the way down to the originating case—patient zero—hence Emile's sobriquet. By asking infected individuals to list all the people they have had contact with during and after the incubation period of the disease—when they are infected but not necessarily symptomatic—scientists can build up a picture of their contact network. By iterating the process many times on the individuals in the network, disease spread can often be pinned down to a single source. As well as allowing us to learn about the complex pattern of disease spread in order to suggest methods to prevent future outbreaks, contact tracing also allows us to take real-time measures to control the spread of disease. It can inform effective strategies to contain a disease in its early stages. Everyone who has had direct contact with an infected individual within the time frame of incubation is quarantined until they have been shown to be free of the disease or to be infected. If infected, they can be kept in isolation until they are no longer likely to pass on the disease.

*   *   *

In practice, though, contact networks are often incomplete, and many disease carriers will not be known to the authorities. Many individuals won't even know that they have the disease due to the incubation period — the window of time, postinfection, before symptoms occur. With Ebola, the incubation period can be up to twenty-one days, but on average it is around twelve days. In October 2014, it became clear that the epidemic in West Africa could potentially take on global proportions. Ostensibly to protect its citizens, the UK government announced that enhanced Ebola screening would take place for passengers entering the UK from high-risk countries at five UK airports and the Eurostar terminal in London.

A similar program in Canada during the time of the SARS (severe acute respiratory syndrome) epidemic in 2004 screened nearly half a million travelers, none of whom were found to have a raised temperature indicative of SARS. The program cost the Canadian government $15 million. In hindsight, the SARS screening program may have helped to reassure the Canadian public that they were safe, but it was ineffective as an intervention strategy.

With that expense in mind, as well as what smacked of a needlessly fraught reaction, a team of mathematicians from the London School of Hygiene & Tropical Medicine developed a simple mathematical model incorporating an incubation period. Considering the average twelve-day incubation period for Ebola and the six-and-a-half-hour flight time from Freetown in Sierra Leone to London, the mathematicians calculated that only about 7 percent of Ebola-carrying individuals boarding planes would be detected by the expensive new measures. They suggested that the money might be better spent on the developing humanitarian crisis in West Africa, which would strike at the source of the problem and consequently reduce the risk of transmission to the UK. This is an example of mathematical intervention at its best — simple, decisive, and evidence based. Rather than speculating on how effective the screening measures could be, a simple mathematical representation of the situation can give us powerful insights and help to direct policy.

R-naught and the Exponential Explosion

The transmission pathway used to identify Emile Ouamouno as Ebola patient zero was far from unique. The disease radiated from its epicenter in Meliandou through multiple distinct pathways. In its early stages, the disease replicated in an exponential fashion through multiple independent channels, much like the memes or viral marketing campaigns described in chapter 1. One person infected three others, who went on to infect others, who infected more still, and so the outbreak exploded in an almost exponential fashion. Whether an outbreak promulgates itself to infamy or dies out into obscurity can be determined by a single number, unique to that outbreak—the basic reproduction number.

Think of a population completely susceptible to a particular disease, much like the inhabitants of Mesoamerica in the 1500s before the arrival of the conquistadores. The average number of previously unexposed individuals infected by a single, freshly introduced disease carrier is known as the basic reproduction number and is often denoted R_0 (pronounced *R-naught* or *R-zero*). If a disease has an R_0 less than one, then the infection will die out quickly as each infectious person passes on the disease, on average, to less than one other individual. The outbreak cannot sustain its own spread. If the R_0 is larger than one, then the outbreak will grow exponentially.

Take a disease such as SARS, for example, with a basic reproduction number of 2. The first person with the disease is patient zero. That person spreads the disease to two others, who each spread the disease to two others, and thence to two others each. Just as we saw in chapter 1, figure 23 (on page 229) illustrates the exponential growth that characterizes the initial phase of the infection. If the spread could continue like this, ten generations down the chain of progression, over a thousand people would be infected. Ten steps further on, the toll would rise to over 1 million.

GENERATION

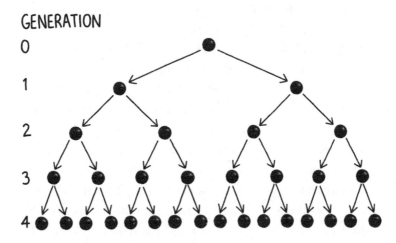

FIGURE 23: Exponential spread of a disease with basic reproduction number, R_0, of 2. The initial infected individual is assumed to be in generation zero. As we enter the fourth generation sixteen people become newly infected.

In practice, just as with the spread of a viral idea, the expansion of a pyramid scheme, the growth of a bacterial colony, or the proliferation of a population, the exponential growth predicted by the basic reproduction number is rarely sustained beyond a few generations. Outbreaks eventually peak and then decline due to the decreasing frequency of infected-susceptible contacts. Ultimately, even when no infectives are left and the outbreak is officially over, some susceptibles will remain. Way back in the 1920s, Kermack and McKendrick came up with a formula that used the basic reproduction number to predict how many susceptible individuals would be left untouched at the end of an outbreak. With an estimated R_0 of around 1.5, Kermack and McKendrick's formula predicts that the Ebola outbreak of 2013–16 would have afflicted 58 percent of the population if no intervention had been taken. In contrast, polio outbreaks have been found to have an R_0 of around 6, which, under Kermack and McKendrick's prediction, means that without interposition only a quarter of 1 percent would survive unscathed.

The basic reproduction number is a ubiquitously useful descriptor of an outbreak because it wraps up all the subtleties of disease transmission into a single figure. From the way in which the infection develops in the body, to the mode of transmission, and even the structure of the societies within which it spreads, it captures all the outbreak's key features and allows us to react accordingly. R_0 can typically be broken down into three components: the size of the population; the rate at which susceptibles become infected (often known as the force of infection); and the rate of recovery or death from the disease. Increasing the first two of these factors increases R_0, while increasing the recovery rate reduces it. The bigger the population and the faster the disease spreads between individuals, the more likely an outbreak is to occur. The quicker individuals recover, the less time they have to pass on the disease to others, and consequently the less likely an outbreak. For many human diseases, we can control only the first two factors. Although antibiotic or antiviral medicines may shorten the course of some diseases, the rate of recovery or fatality is often a property of the disease-causing pathogen. A quantity closely related to R_0 is the *effective* reproduction number (often denoted R_e)—the average number of secondary infections caused by an infectious individual at a *given point* in the outbreak's progression. If, by intervention, R_e can be brought below one, then the disease will die out.

Although crucial for disease control, R_0 does not tell us how serious a disease is for an infected individual. An extremely infectious disease such as measles, for example, with an R_0 of between 12 and 18, is typically considered less serious for an individual than a disease such as Ebola, with a smaller R_0 of around 1.5. While measles spreads quickly, its case fatality rate is small compared to the 50–70 percent of Ebola patients who will die of the disease.

Perhaps surprisingly, diseases with high case fatality rates tend to be less infectious. If a disease kills too many of its victims too quickly, then it reduces its chances of being passed on. Diseases that kill most of the people they infect and also spread efficiently are very rare and are usually confined to disaster movies. Although a high case fatality rate significantly raises the fear associated with an outbreak, diseases with high R_0 but lower case fatality may end up killing more people by virtue of the larger numbers they infect.

The math dictates that once we have decided that a disease needs to be controlled, case fatality rates don't provide useful information on how to decelerate the spread. The three factors that comprise R_0, however, suggest important interventions that can bring a halt to deadly disease outbreaks before they run their unencumbered courses.

Taking Control

One of the most effective options for reducing disease spread is vaccination. By taking people directly from susceptible to removed, bypassing the infective state, it effectively reduces the size of the susceptible population. Vaccination, however, is typically a precautionary measure applied in an attempt to reduce the probability of outbreaks. Once outbreaks are in full swing, it is often impractical to develop and test an effective vaccine in time.

An alternative strategy, employed for animal diseases, that has the same diminishing effect on R_e, the effective reproduction number, is culling. In 2001, when Britain was in the grip of the foot-and-mouth crisis, the decision was taken to cull. By slaughtering infected individuals the infectious period was lowered from up to three weeks to a matter of days, dramatically reducing the effective reproduction number. In this outbreak, however, culling only the infected animals was not enough to control the disease. Some infectives inevitably slipped the net, causing infection in others nearby. In response, the government implemented a "ring culling" strategy, slaughtering animals (infected or not) within a three-kilometer radius of affected farms. At first glance killing uninfected individuals seems a pointless exercise. However, because it reduces the population of susceptible animals in the local area—one of the factors that contributes to the reproduction number—the math dictates that it slows the spread of the disease.

For active outbreaks of human diseases in unvaccinated populations, culling is clearly not an option. Quarantine and isolation, however, can be extremely efficient ways to reduce the transmission rate and, consequently, the effective reproduction number. Isolating infective patients reduces the rate of spread, while quarantining healthy individuals reduces the effective susceptible population. Both actions contribute to decreasing the effective reproduction number. Indeed, the last smallpox outbreak in Europe, in Yugoslavia in 1972, was rapidly brought under control by extreme quarantine measures. Up to ten thousand potentially infectious individuals were held under armed guard in hotels commandeered for that express purpose, until the threat of new cases had passed.

In less extreme cases, simple applications of mathematical modeling can suggest the most effective duration to isolate infected patients. They can also determine whether to quarantine a proportion of the uninfected population, weighing up the economic costs of quarantining healthy individuals against the risk of an enlarged disease outbreak. This sort of mathematical modeling really comes into its own for situations in which carrying out field studies on disease progression is impractical for logistical or ethical reasons. For example, it's inhumane, during a disease outbreak, to deprive for the purposes of a study a fraction of a population of an intervention that might save lives. Similarly, it's impractical in the real world to quarantine a high proportion of the population for a long time. Running a mathematical model presents no such concerns. We can test models in which everyone is quarantined or no one, or anywhere in between, in an attempt to balance the economic impact of this enforced isolation with the effect it has on the progression of the disease.

This is the real beauty of mathematical epidemiology—the ability to test out scenarios that are infeasible in the real world, sometimes with surprising and counterintuitive results. Math has, for example, shown that for diseases such as chicken pox (varicella) isolation and quarantine may be the wrong strategy. Trying to segregate children with and without the disease will undoubtedly lead to numerous missed schooldays and workdays to avoid what is widely considered to be a relatively mild disease. Perhaps more significant, though, mathematical models prove that quarantining healthy children can defer their catching the disease until they are older, when the complications from chicken pox can be far more serious. Such counterintuitive effects of a seemingly sensible strategy such as isolation might never have been fully understood if not for mathematical interventions.

If quarantine and isolation have unexpected consequences for some diseases, they simply don't work at all for others. Mathematical models of disease spread have identified that the degree to which a quarantining strategy succeeds depends on the *timing* of peak infectiousness. If a disease is especially infectious in the early stages, when patients are asymptomatic, then they may spread the disease to the majority of their expected victims before they can be isolated. Fortunately, in the case of Ebola, for which many other potential control avenues are blocked, the majority of transmissions occur after patients are symptomatic and can be isolated.

The infectious period of Ebola extends to such an extreme that—even after their death—victims' viral loads remain high. The dead may still infect individuals who come into contact with their corpses. Notably, the funeral of one traditional healer in Sierra Leone was one of the major flash points in the early spread of the outbreak. With the cases increasing rapidly throughout Guinea, people started to become more desperate. Apprised of the well-renowned healer's powers, Ebola patients from Guinea crossed the border into Sierra Leone to consult her. She believed she could cure the disease, but unsurprisingly, she quickly fell sick herself and died. Her burial attracted hundreds of mourners over several days, all observing traditional funeral practices including washing and touching the corpse. That single event was directly linked to over 350 Ebola deaths and facilitated the full-blown introduction of the disease into Sierra Leone.

In 2014, at around the peak of the Ebola outbreak, a mathematical study concluded that approximately 22 percent of new Ebola cases were attributable to deceased Ebola victims. The same study suggested that by limiting traditional practices, including burial rituals, the basic reproduction number might be so reduced that the outbreak would become unsustainable. One of the most important interventions enforced by the governments of West Africa and humanitarian organizations working in the area was to restrict traditional funeral procedures and ensure that all Ebola victims were given safe and dignified burials. In combination with education campaigns providing alternatives to unsafe traditional practices and imposing travel restrictions even on seemingly healthy individuals, the Ebola outbreak was eventually brought to heel. On June 9, 2016, some two and a half years after Emile Ouamouno's infection, the West African Ebola outbreak was declared over.

Herd Immunity

As well as actively helping to tackle infectious disease, mathematical models of epidemics can also help us to understand the unusual features of different disease landscapes. For example, a number of interesting questions surround childhood diseases such as mumps and rubella: Why do these diseases sweep over us periodically, affecting only children? Perhaps they have a particular predilection toward some elusive childhood quality? And why have they persisted for so long in our society? Perhaps they lie dormant for several years, biding their time between major outbreaks only to strike our most defenseless?

Childhood diseases show these typical periodic outbreak patterns because the effective reproduction number varies over time with the population of susceptible individuals. After a big outbreak has affected large swathes of the unprotected-child population, a disease such as scarlet fever doesn't just disappear. It persists in the population, but with an effective reproduction number that hovers around 1. The disease only just sustains itself. As time goes by, the population ages, and new, unprotected children are born. As the unguarded fraction of the population grows, the effective reproduction number becomes higher and higher, making new outbreaks increasingly likely. When an outbreak finally takes off, the victims to whom the disease spreads are usually at the unprotected younger end of the demographic, because most of the older populace are already immune through experiencing the disease. Those people who didn't get the disease as children are typically afforded some protection because they fraternize with fewer of the infected age group.

The idea that a large population of immune individuals can slow or even halt the spread of a disease, as with the dormant periods between outbreaks of childhood diseases, is a mathematical concept known as herd immunity. Surprisingly, this community effect does not require everyone to be immune to the disease for the whole population to be protected. By reducing the effective reproduction number to less than one, the chain of transmission can be broken and the disease stopped in its tracks. Crucially, herd immunity means that those with immune systems too weak to tolerate vaccination, including the elderly, newborns, pregnant women, and people with HIV, can still benefit from the protection of vaccinations. What fraction of the population needs to be immune to protect the susceptible portion varies depending on how infectious the disease is. The basic reproduction number, R_0, holds the key to how large that proportion is.

Take, for example, a person infected with a virulent strain of flu illustrated in figure 24. If the person meets 20 susceptible people during the week in which they are infectious and 4 of them become infected, then the basic reproduction number of the disease, R_0, is 4. Each susceptible person has a one-in-five chance of being infected. This illustrates how the reproduction number depends on the size of the susceptible population. If our flu patient had only met 10 susceptible people during their infectious week (as in the middle panel of figure 24), with the probability of transmission remaining the same, then they would have infected only two of them, on average, halving the effective reproduction number from 4 to 2.

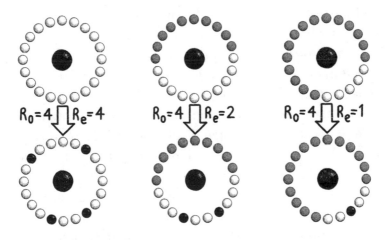

FIGURE 24: A single infectious individual (*black*) meets 20 susceptible (*white*) or vaccinated (*gray*) individuals during her week-long infectious period. When no one is vaccinated (*left*) the single infectious individual infects 4 others, meaning the basic reproduction number, R_0, is 4. When half of the population is vaccinated (*middle*) only 2 susceptible individuals become infected. The effective reproduction number, R_e, is reduced to 2. Finally (*right*) when ¾ of the population is vaccinated, only 1 other person becomes infected on average. The effective reproduction number is reduced to the critical value of 1.

The most effective way to reduce the size of the susceptible population is through vaccination. The question of how many to vaccinate to achieve herd immunity relies on reducing the effective reproduction number to below 1. If we could vaccinate ¾ of the population, then (as in the situation on the right of figure 24), of the original 20 contacts our flu patient made in a week, only ¼ (i.e., 5) would still be susceptible. On average, only 1 of them would become infected. It's no coincidence that this critical vaccination threshold, for achieving herd immunity for a disease with basic reproduction number 4, requires ¾ (which is 1–¼) of the population to be vaccinated. In general we can only afford to leave $1/R_0$ of the population unvaccinated and must protect the remaining fraction ($1–1/R_0$ of the population) if we are to achieve the herd immunity threshold. For smallpox, whose basic reproduction number is around 4, we can afford to leave ¼ (or 25 percent) of the population unprotected. Vaccinating only 80 percent (to provide a buffer, 5 percent above the 75 percent critical immunization threshold) of the susceptible population against smallpox was enough, in 1977, to complete one of the greatest accomplishments of our species—to wipe a human disease clean off the face of the Earth. The feat has never been repeated.

The debilitating and dangerous implications of smallpox infection alone made it a suitable target for eradication. Its low critical immunization threshold also made it a relatively easy target. Many diseases are harder to protect against because they spread more easily. Chicken pox, with an estimated R_0 of around 10, would require ⁹⁄10 of the population to be immune before the rest would be effectively protected and the disease wiped out. Measles, by far the most infectious human-to-human disease on Earth, with an R_0 estimated to be between 12 and 18, would require between 92 percent and 95 percent of the population to be vaccinated. A study that modeled the spread of the 2015 Disneyland measles outbreak—in which Mobius Loop was infected—suggested vaccination rates among those exposed to the disease may have been as low as 50 percent, way below the threshold required for herd immunity.

Mr. MMR

Since its introduction in 1988, England's rate of vaccination against measles, through the combined measles, mumps, and rubella (MMR) injection, had been steadily rising. In 1996, the vaccination rate hit a record high at 91.8 percent—close to the critical immunization threshold for eliminating measles. Then, in 1998, something happened that would derail the vaccination process for years.

This public health disaster was not caused by disease-ridden animals, poor sanitation, or even failures of government policy, but instead by a somber five-page publication in the well-respected medical journal the *Lancet*. In the study, lead author Andrew Wakefield proposed a link between the MMR vaccine and autism spectrum disorders. On the back of his "findings," Wakefield launched his own personal anti-MMR campaign, stating in a press conference, "I can't support the continued use of these three vaccines given in combination until this issue has been resolved." Most of the mainstream media couldn't resist the bait.

Among the *Daily Mail*'s headlines covering the story were "MMR Killed My Daughter," "MMR Fears Gain Support," and "MMR Safe? Baloney. This Is One Scandal That's Getting Worse." In the years that followed Wakefield's article, the story snowballed, becoming the biggest UK science story of 2002. While indulging the fears of many fretful parents, the media's coverage of the story typically failed to mention that Wakefield's study was conducted on just twelve children, an extremely small cohort from which to draw meaningful large-scale conclusions. Any coverage that did sound a note of caution about the study was drowned out by the warning sirens emanating from most news outlets. As a result, parents started to withdraw permission for their children to be vaccinated. In the ten years that followed the publication of the infamous *Lancet* paper, the MMR uptake rate would drop from above 90 percent to below 80 percent. Confirmed cases of measles would increase from fifty-six in 1998 to over thirteen hundred ten years later.

Cases of mumps, which had been becoming less prevalent throughout the 1990s, suddenly skyrocketed.

In 2004, as instances of measles, mumps, and rubella continued to increase, one investigative journalist, Brian Deer, sought to expose Wakefield's work as fraudulent. Deer reported that, prior to submitting his paper, Wakefield had received over £400,000 from lawyers looking for evidence against the pharmaceutical companies that manufacture vaccines. Deer also uncovered documents that, he claimed, showed that Wakefield had submitted patents for a rival vaccine to MMR. Crucially, Deer claimed to have evidence that Wakefield had manipulated the data in his paper to give the false impression of a link to autism. Deer's evidence of Wakefield's scientific fraud and extreme conflicts of interest eventually led to the offending paper's retraction by the *Lancet*'s editors. In 2010, Wakefield was struck off the medical register by the General Medical Council. In the twenty years since Wakefield's original paper, at least fourteen comprehensive studies on hundreds of thousands of children across the world have found no evidence of a link between MMR and autism. Sadly, though, Wakefield's influence lives on.

Although MMR vaccination in the UK has returned to pre-scare levels, vaccination rates across the developed world as a whole are dropping, and measles cases are increasing. In Europe, 2018 saw more than sixty thousand cases of measles, with seventy-two proving fatal—double the number from the previous year. The United States experienced more measles cases in the first four months of 2019 than in any year for a quarter of a century. In the main this is a result of the emergence of the growing anti-vaccination movement. The World Health Organization lists what it calls "vaccine hesitancy" as one of 2019's top ten global health threats. The *Washington Post*, among other media outlets, attributes the rise of the "anti-vaxxers" directly to Wakefield, describing him as "the founder of the modern anti-vaccination movement." The doctrines of the movement, however, have expanded far beyond Wakefield's now-debunked findings. They range from assertions that vaccines contain dangerous levels of toxic chemicals to allegations that vaccines actually infect children with the diseases they are trying to prevent. In reality, toxic chemicals such as formaldehyde are produced in higher amounts by our own metabolic system than the trace amounts found in vaccines. Similarly, vaccinations cause the disease they are designed to protect against extremely rarely, especially so in otherwise healthy individuals.

Despite many convincing refutations of the anti-vaxxers' claims, their rhetoric has risen to prominence as a result of support from high-profile celebrities including Jim Carrey, Charlie Sheen, and Donald Trump. In a twist almost too implausible to be believed, in 2018, Wakefield confirmed his own elevation to celebrity status when he started dating former supermodel Elle Macpherson.

Alongside the rise of the celebrity activist has come the emergence of social media, allowing these personalities to promulgate their views directly to their fans on their own terms. With the erosion of trust in the mainstream media, people are increasingly turning to these echo chambers for reassurance. The rise of these alternative platforms has provided a space for the anti-vaccination movement to grow unthreatened and unchallenged by evidence-based science. Wakefield himself even described the emergence of social media as having "evolved beautifully" — for his purposes, perhaps.

We all have choices to make that affect our likelihood of contracting infectious disease: whether to holiday in exotic countries; whom to let our children play with; whether we travel on crowded public transport. When we are ill, other choices we make affect our likelihood of transmitting disease to others: whether we cancel the much-anticipated catch-up with our friends; whether we keep our children home from school; whether we cover our mouths when we cough. The crucial decision on whether we vaccinate ourselves and our dependents can only be taken ahead of time. It affects our chances not only of catching but also of transmitting diseases.

Some of these decisions are inexpensive, making their adoption straightforward. It costs nothing to sneeze into a tissue or a handkerchief.

Simply washing your hands frequently and carefully has been shown to reduce the effective reproduction numbers of respiratory illnesses such as flu by as much as three-quarters. For some diseases, this might be enough to take us below the threshold value of R_0, so that an infectious disease cannot break out.

Other decisions provide us with more of a dilemma. It is always tempting to send the kids to school even if we know it increases the number of potentially infectious contacts they will make and thus increases the possibility of an epidemic. At the heart of all our choices should be an understanding of their risks and consequences.

Mathematical epidemiology provides a way to assess and understand these decisions. It explains why it is better for everyone if you stay away from work or school if you are ill. It tells us how and why washing our hands can help prevent outbreaks by reducing the force of infection. Sometimes, counterintuitively, it can highlight that the most terror-inducing diseases are not always the ones we should worry about most.

On a broader scale it suggests strategies to tackle disease outbreaks and the preventive measures we can take to avoid them. In conjunction with reliable scientific evidence, mathematical epidemiology demonstrates that vaccination is a no-brainer. Not only does it protect you, it protects your family, your friends, your neighbors, and your colleagues. World Health Organization figures show that vaccines prevent millions of deaths every year and could prevent millions more if we could improve global coverage. They are the best way we have of preventing outbreaks of deadly diseases and the only chance we have of terminating their devastating impacts for good. Mathematical epidemiology is a glimmer of hope for the future, a key that can unlock the secrets of how to achieve these monumental tasks.

MATHEMATICAL EMANCIPATION

Mathematics has shaped our history, through the ancestors who won evolution's numbers game and the diseases that strained and filtered our species. Our biology reflects mathematics' constant unchanging rules. At the same time, our mathematical aesthetics have morphed to reflect our very physiology, and our mathematical understanding has coevolved with us over millions of years to its current state.

In today's society, mathematics underpins almost everything we do. It is vital to the ways in which we communicate with one another and the methods we use to navigate from place to place. It has completely altered how we buy and sell, and it has revolutionized the manner in which we work and relax. Its influence can be felt in almost every courtroom and every hospital ward, in every office and every home.

Math is being used daily to achieve previously unimaginable tasks. Sophisticated algorithms allow us to find the answer to almost any question in seconds. People across the world are linked in an instant by the mathematical power of the internet. The guardians of justice use mathematics as a force for good when detecting criminals through forensic archaeology.

We must remember, however, that mathematics is only as benign as the person or people who wield it. The same mathematics that implicated the art forger Han van Meegeren also gave us the atomic bomb. Clearly we should strive to understand the full implications of the mathematical tools we so frequently submit ourselves to. What starts with friend recommendations and personalized advertising might end with trending fake news or the erosion of our privacy.

As mathematics becomes an increasingly prevalent part of our everyday lives, the opportunities for unexpected disaster multiply. As many times as we have appreciated the wonderful uses of mathematics to achieve hitherto unthinkable feats, we have seen the disastrous consequences of mathematical mistakes. Careful mathematics may have put man on the moon, but careless mathematics destroyed the multimillion-dollar Mars Orbiter. When handled appropriately, math can be a powerful tool for criminal analysis, but when abused by unscrupulous shysters it can cost innocents their liberty. At its best, math is the state-of-the-art medical technology that can save lives, but at its worst it is the miscalculated doses that end them. It is our duty to learn from mathematical mistakes so they are not revisited in the future or, better still, are rendered completely unrepeatable.

Mathematical modeling can provide us with some access to what that future will look like. Mathematical models don't just describe the world as it is—the data they are calibrated against—but rather they provide some degree of clairvoyance. Mathematical epidemiology allows us to peer into the future of disease progression and to take proactive preventive measures, rather than always playing reactive games of catch-up. Optimal stopping can give us the best chance of making the best choice when we are not allowed to see all the options beforehand. Personal genomics may revolutionize the understanding of our future disease risk, but only if we can standardize the mathematics with which we interpret the results.

Mathematics has been, is, and ever will be a current running almost unseen beneath the surface of our affairs. We should, however, be careful not to get carried away with the flow by attempting to extend its application beyond its remit. In some places mathematics is completely the wrong tool for the job, where human supervision is unquestionably necessary. Even if some of the most complex mental tasks can be farmed out to an algorithm, matters of the heart can never be broken down into a simple set of rules. No code or equation will ever imitate the true complexities of the human condition.

Nevertheless, a little mathematical knowledge in our increasingly quantitative society can help us to harness the power of numbers for ourselves. Simple rules allow us to make the best choices and avoid the worst mistakes. Small alterations in the way we think about our rapidly evolving environments help us to "keep calm" in the face of rapidly accelerating change, or to adapt to our increasingly automated realities. Basic models of our actions, reactions, and interactions can prepare us for the future before it arrives. The stories relating other people's experiences are, in my view, the simplest and most powerful models of all. They allow us to learn from the mistakes of our predecessors so that, before we embark on any numerical expedition, we ensure we are all speaking the same language, have synchronized our watches, and checked we've got enough fuel in the tank.

Half the battle for mathematical empowerment is daring to question the perceived authority of those who wield the weapons—shattering the illusion of certainty. Appreciating absolute and relative risks, ratio biases, mismatched framing, and sampling bias gives us the power to be skeptical of the statistics screamed from newspaper headlines, the "studies" pushed at us in ads, or the half-truths that come tumbling from the mouths of our politicians. Recognizing ecological fallacies and dependent events allows us to disperse obfuscating smoke screens, making it harder to fool us with mathematical arguments, be they in the courtroom, the classroom, or the clinic.

We must ensure that the person with the most shocking statistics doesn't always win the argument, by demanding an explanation of the math behind the figures. We shouldn't let medical charlatans delay us from receiving potentially lifesaving treatment when their alternative therapies are just a regression to the mean. We mustn't let the anti-vaxxers make us doubt the efficacy of vaccinations when mathematics proves that they can save vulnerable lives and wipe out disease.

It is time for us to take the power back into our own hands, because sometimes math really is a matter of life and death.

ACKNOWLEDGMENTS

The title and the main thrust for *The Math of Life and Death*, as a book about the hidden places in which math affects our everyday lives, came from a drunken conversation in the pub the first time I met my agent, Chris Wellbelove, face-to-face. Chris has looked at every draft of every pitch and chapter that I have sent him and done so much more besides. I owe him a great debt for taking a chance on me and for steering me successfully through the process of pitching and writing my first book.

From the day I signed with Quercus, my editor, Katy Follain, has had my back. She has looked at numerous drafts of the book and made suggestions that have immeasurably improved it. Similarly, Sarah Goldberg, my US editor, has had a huge influence on the direction that the book has taken. That Katy and Sarah took the time to get their heads together and give me coherent feedback makes me all the more grateful. Thanks to them and to all the other people who have worked tirelessly behind the scenes at Quercus and Scribner to make this book happen.

I owe a huge debt of gratitude to all the people whom I contacted when writing this book and who so kindly agreed to share their stories with me. Your tales of mathematical catastrophes and triumphs are the fabric of the book. It simply wouldn't have happened without the time and generosity you put in answering my long lists of seemingly irrelevant questions.

I am grateful to the Institute for Mathematical Innovation at the University of Bath, which has supported me through an internal secondment. This has been hugely influential in allowing me to deliver the book I wanted with the care it deserved. More broadly, many of my colleagues across the wider university to whom I have spoken about the book have given me great encouragement and support. My alma mater, Somerville College, Oxford, also provided me a place to work when I needed to be out of the house, and I am grateful for that.

At the start of the writing process, when I realized I needed sound quantitative minds to critique my work, I looked to my former PhD colleagues and close friends Gabriel Rosser and Aaron Smith. Not quite knowing what they were letting themselves in for, they agreed to look over early drafts of the manuscript despite both having small babies and many other life complications to contend with. My sincerest gratitude goes to them for the improvements their comments have made to the book.

My great friend and colleague Chris Guiver has been so kind as to let me lodge at his house once a week for over a year while I have been writing this book. He has acted as an excellent sounding board for my ideas, discussing and debating with me long into the night about the book, about science, and about life more generally. Chris, you probably don't realize how much your generosity has meant to me. Thank you.

My parents, Tim and Mary, have been my most staunch supporters throughout this process. They have read the whole book through *twice*. They are my audience of intelligent laypeople. More than just their insightful comments and thorough proofreading, however, I owe them for my education and my values. You have supported me through highs and lows. I will never be able to thank you enough.

My sister, Lucy, helped me to weave the threads of my early ideas together into something resembling a coherent pitch. It is no exaggeration to say the book would not exist without the time and effort she put into sensitively critiquing my writing and setting me on the right path at the start.

I owe a, perhaps less tangible, debt of thanks to my wider family for, among other things, never complaining when I went off to sleep in the middle of a family gathering because I'd stayed up late the previous night to work on the book. The importance of that respite cannot be overestimated.

Finally, to the people who have probably endured most for this book: my family. My wife, Caroline, has been incredibly supportive of the project, even fiddling with the genetics sections of the book when I let her. Not only has she been supporting a fledgling author, but she's also been a badass mum and full-time CEO to boot. My admiration for you is unwavering. Lastly, to Em and Will. Thanks for keeping my feet on the ground. Every worry goes out of my head when I come home; there isn't space for anything but you two. Even if this book doesn't sell a single copy, I know it won't make a difference to you.

ACKNOWLEDGMENTS

REFERENCES

Introduction: Almost Everything

3 *By providing an estimate of the number of fish in a lake:* K. H. Pollock, "Modeling Capture, Recapture, and Removal Statistics for Estimation of Demographic Parameters for Fish and Wildlife Populations: Past, Present, and Future," *Journal of the American Statistical Association* 86 (413) (1991): 225, https://doi.org/10.2307/2289733.

3 *the number of drug addicts:* M. L. Doscher and J. A. Woodward, "Estimating the Size of Subpopulations of Heroin Users: Applications of Log-Linear Models to Capture/Recapture Sampling," *International Journal of the Addictions* 18 (2) (1983): 167–82; R. Hartnoll, M. Mitcheson, R. Lewis, and S. Bryer, "Estimating the Prevalence of Opioid Dependence," *Lancet* 325 (8422) (1985): 203–5, https://doi.org/10.1016/S0140-6736(85)92036-7; J. A. Woodward, R. L. Retka, and L. Ng, "Construct Validity of Heroin Abuse Estimators," *International Journal of the Addictions* 19 (1) (1984): 93–117, https://doi.org/10.3109/10826088409055819.

3 *the number of war dead in Kosovo:* M. Spagat, "Estimating the Human Costs of War: The Sample Survey Approach," in *The Oxford Handbook of the Economics of Peace and Conflict,* ed. Michelle R. Garfinkel and Stergios Skaperdas (Oxford, UK: Oxford University Press, 2012), https://doi.org/10.1093/oxfordhb/9780195392777.013.0014.

Chapter 1: Thinking Exponentially

10 Strep f. *is one of the bacteria responsible for the souring:* S. G. Botina, A. M. Lysenko, and V. V. Sukhodolets, "Elucidation of the Taxonomic Status of Industrial Strains of Thermophilic Lactic Acid Bacteria by Sequencing of 16S rRNA Genes," *Microbiology* 74 (4) (2005): 448–52, https://doi.org/10.1007/s11021-005-0087-7.

10 Strep f. *cells can divide to produce two daughter cells every hour:* A. M. Cárdenas, K. A. Andreacchio, and P. H. Edelstein, "Prevalence and Detection of

Mixed-Population Enterococcal Bacteremia," *Journal of Clinical Microbiology* 52 (7) (2014): 2604–8, https://doi.org/10.1128/JCM.00802-14; M. M. C. Lam et al., "Comparative Analysis of the Complete Genome of an Epidemic Hospital Sequence Type 203 Clone of Vancomycin-Resistant *Enterococcus faecium*," *BMC Genomics* 14 (2013): 595, https://doi.org/10.1186/1471-2164-14-595.

17 *He published in the journal* Nature *evidence:* H. Von Halban, F. Joliot, and L. Kowarski, "Number of Neutrons Liberated in the Nuclear Fission of Uranium," *Nature* 143 (3625) (1939): 680, https://doi.org/10.1038/143680a0.

18 *(the other significant isotope, which makes up 99.3 percent of naturally occurring uranium):* J. Webb, "Are the Laws of Nature Changing with Time?," *Physics World* 16 (4) (2003): 33–38, https://doi.org/10.1088/2058-7058/16/4/38.

21 *One kilogram of U-235 can release roughly 3 million:* J. Bernstein, *Nuclear Weapons: What You Need to Know* (Cambridge, UK: Cambridge University Press, 2008).

23 *The fire threw into the atmosphere hundreds of times more radioactive material:* International Atomic Energy Agency, "Ten Years after Chernobyl: What Do We Really Know?," in *Proceedings of the IAEA/WHO/EC International Conference: One Decade after Chernobyl: Summing Up the Consequences* (Vienna: International Atomic Energy Agency, 1996).

24 *the elimination of drugs in the body:* D. J. Greenblatt, "Elimination Half-Life of Drugs: Value and Limitations," *Annual Review of Medicine* 36 (1) (1985): 421–27, https://doi.org/10.1146/annurev.me.36.020185.002225; I. M. Hastings, W. M. Watkins, and N. J. White, "The Evolution of Drug-Resistant Malaria: The Role of Drug Elimination Half-Life," *Philosophical Transactions of the Royal Society of London, Series B: Biological Sciences* 357 (1420) (2002): 505–19, https://doi.org/10.1098/rstb.2001.1036.

24 *the rate of decrease of the head on a pint of beer:* A. Leike, "Demonstration of the Exponential Decay Law Using Beer Froth," *European Journal of Physics* 23 (1) (2002): 21–26, https://doi.org/10.1088/0143-0807/23/1/304; N. Fisher, "The Physics of Your Pint: Head of Beer Exhibits Exponential Decay," *Physics Education* 39 (1) (2004); 34–35, https://doi.org/10.1088/0031-9120/39/1/F11.

24 *the rate at which the levels of radiation emitted by a radioactive substance decrease:* E. Rutherford and F. Soddy, "LXIV. The Cause and Nature of Radioactivity—Part II," *London, Edinburgh, and Dublin Philosophical Magazine and Journal of Science* 4 (23) (1902): 569–85, https://doi.org/10.1080/14786440209462881d; E. Rutherford and F. Soddy, "XLI. The Cause and Nature of Radioactivity—Part I," *London, Edinburgh, and Dublin Philosophical Magazine and Journal of Science,* 4 (21) (1902): 370–96, https://doi.org/10.1080/14786440209462856.

25 *determining the age of ancient artifacts such as the Dead Sea Scrolls:* G. Bonani et al., "Radiocarbon Dating of Fourteen Dead Sea Scrolls," *Radiocarbon* 34

(3) (1992): 843–49, https://doi.org/10.1017/S0033822200064158; I. Carmi, "Radiocarbon Dating of the Dead Sea Scrolls," in *The Dead Sea Scrolls: Fifty Years after Their Discovery, 1947–1997*, ed. L. Schiffman, E. Tov, and J. VanderKam (Jerusalem: Israel Exploraton Society, 2000), 881; G. Bonani, M. Broshi, and I. Carmi, "14 Radiocarbon Dating of the Dead Sea Scrolls," '*Atiqot* (Israel Antiquities Authority, 1991).

25 *archaeopteryx was 150 million years old:* C. Starr, R. Taggart, C. A. Evers, and L. Starr, *Biology: The Unity and Diversity of Life* (Boston: Cengage Learning, 2019).

25 *Ötzi the iceman died fifty-three hundred years ago:* G. Bonani et al., "Ams 14C Age Determinations of Tissue, Bone and Grass Samples from the Ötztal Ice Man," *Radiocarbon* 36 (2) (1994): 247–50, https://doi.org/10.1017/S0033822200040534.

27 *This established for certain that Van Meegeren's forgeries:* B. Keisch, R. L. Feller, A. S. Levine, and R. R. Edwards, "Dating and Authenticating Works of Art by Measurement of Natural Alpha Emitters," *Science* 155 (3767) (1967): 1238–42, https://doi.org/10.1126/science.155.3767.1238.

29 *researchers discovered a third gene responsible for ALS:* K. P. Kenna et al., "NEK1 Variants Confer Susceptibility to Amyotrophic Lateral Sclerosis," *Nature Genetics* 48 (9) (2016): 1037–42, https://doi.org/10.1038/ng.3626.

30 *Computer scientist Vernor Vinge encapsulated just such ideas:* V. Vinge, *Marooned in Realtime* (New York: Bluejay Books / St. Martin's Press, 1986), *A Fire upon the Deep* (New York: Tor Books, 1992), and *The Coming Technological Singularity: How to Survive in the Post-Human Era*, in NASA, *Lewis Research Center, Vision 21: Interdisciplinary Science and Engineering in the Era of Cyberspace* (1993), 11–22, https://ntrs.nasa.gov/search.jsp?R=19940022856.

30 *Kurzweil hypothesized "the law of accelerating returns":* R. Kurzweil, *The Age of Spiritual Machines: When Computers Exceed Human Intelligence* (New York: Viking, 1999).

30 *"technological change so rapid and profound":* R. Kurzweil, "The Law of Accelerating Returns," in *Alan Turing: Life and Legacy of a Great Thinker* (Berlin and Heidelberg: Springer Berlin Heidelberg, 2004), 381–416, https://doi.org/10.1007/978-3-662-05642-4_16.

31 *The complete "Book of Life" was delivered in 2003:* S. G. Gregory et al., "The DNA Sequence and Biological Annotation of Human Chromosome 1," *Nature* 441 (7091) (2006): 315–21, https://doi.org/10.1038/nature04727; International Human Genome Sequencing Consortium, "Initial Sequencing and Analysis of the Human Genome," *Nature* 409 (6822) (2001): 860–921, https://doi.org/10.1038/35057062; E. Pennisi, "The Human Genome," *Science* 291 (5507) (2001): 1177–80, https://doi.org/10.1126/SCIENCE.291.5507.1177.

32 *human population grows at a rate that is proportional to its current size:*

T. R. Malthus and G. Gilbert, *An Essay on the Principle of Population* (Oxford, UK: Oxford University Press, 2008).

33 *the first to demonstrate that logistic growth occurred in bacterial populations*: A. G. McKendrick and M. K. Pai, "The Rate of Multiplication of Microorganisms: A Mathematical Study," *Proceedings of the Royal Society of Edinburgh* 31 (1912): 649–53, https://doi.org/10.1017/S0370164600025426.

33 *sheep*: J. Davidson, "On the Ecology of the Growth of the Sheep Population in South Australia," *Transactions of the Royal Society of South Australia* 62 (1) (1938): 11–148; J. Davidson, "On the Growth of the Sheep Population in Tasmania," *Transactions of the Royal Society of South Australia* 62 (2) (1938): 342–46.

33 *seals*: S. Jeffries, H. Huber, J. Calambokidis, and J. Laake, "Trends and Status of Harbor Seals in Washington State: 1978–1999," *Journal of Wildlife Management* 67 (1) (2003): 207, https://doi.org/10.2307/3803076.

33 *cranes*: M. N. Flynn and W. R. L. S. Pereira, "Ecotoxicology and Environmental Contamination," *Ecotoxicology and Environmental Contamination* 8 (1) (2013): 75–85.

34 *size of human population that Earth's biosphere can support*: E. O. Wilson, *The Future of Life* (New York: Alfred A. Knopf, 2002).

34 *projected to fall below 1 percent per year by 2023*: A. E. Raftery, L. Alkema, and P. Gerland, "Bayesian Population Projections for the United Nations," *Statistical Science: A Review Journal of the Institute of Mathematical Statistics* 29 (1) (2014): 58–68, https://doi.org/10.1214/13-STS419; A. E. Raftery et al., "Bayesian Probabilistic Population Projections for All Countries," *Proceedings of the National Academy of Sciences of the United States of America* 109 (35) (2012): 13915–21, https://doi.org/10.1073/pnas.1211452109; United Nations Department of Economic and Social Affairs Population Division, "World Population Prospects: The 2017 Revision, Key Findings and Advance Tables," ESA/P/WP/2, 2017.

35 *perceived time really does seem to run more quickly the older we get*: R. A. Block, D. Zakay, and P. A. Hancock, "Developmental Changes in Human Duration Judgments: A Meta-analytic Review," *Developmental Review* 19 (1) (1999): 183–211, https://doi.org/10.1006/DREV.1998.0475.

35 *older group didn't call a halt until a staggering three minutes and forty seconds*: P. Mangan, P. Bolinskey, and A. Rutherford, "Underestimation of Time during Aging: The Result of Age-Related Dopaminergic Changes" (Annual Meeting of the Society for Neuroscience, 1997).

35 *Older participants consistently gave shorter estimates*: F. I. M. Craik and J. F. Hay, "Aging and Judgments of Duration: Effects of Task Complexity and Method of Estimation," *Perception and Psychophysics* 61 (3) (1999): 549–60, https://doi.org/10.3758/BF03211972.

36 *our metabolism slows as we get older*: R. M. Church, "Properties of the Internal

Clock," *Annals of the New York Academy of Sciences* 423 (1) (1984): 566–82, https://doi.org/10.1111/j.1749-6632.1984.tb23459.x; Craik and Hay, "Aging and Judgments of Duration," 549–60; J. Gibbon, R. M. Church, and W. H. Meck, "Scalar Timing in Memory," *Annals of the New York Academy of Sciences* 423 (1984): 52–77, https://doi.org/10.1111/j.1749-6632.1984.tb23417.x.

36 *our perception of time's passage depends upon the amount of new perceptual information:* Pennisi, "Human Genome," 1177–80.

36 *Experiments on subjects experiencing the unfamiliar sensation of free fall:* C. Stetson, M. P. Fiesta, and D. M. Eagleman, "Does Time Really Slow Down during a Frightening Event?," *PLoS ONE* 2 (12) (2007): e1295, https://doi .org/10.1371/journal.pone.0001295.

Chapter 2: Sensitivity, Specificity, and Second Opinions

46 *using the same method as 23andMe and data taken directly from the report:* L. A. Farrer et al., "Effects of Age, Sex, and Ethnicity on the Association between Apolipoprotein E Genotype and Alzheimer Disease," *JAMA* 16 (1997): 1349, https://doi.org/10.1001/jama.1997.03550160069041; J. Gaugler et al., "2016 Alzheimer's Disease Facts and Figures," *Alzheimer's & Dementia* 12 (4) (2016): 459–509, https://doi.org/10.1016/J.JALZ.2016.03.001; E. Genin et al., "APOE and Alzheimer Disease: A Major Gene with Semi-dominant Inheritance," *Molecular Psychiatry* 16 (9) (2011): 903–7, https://doi .org/10.1038/mp.2011.52; N. P. Jewell, *Statistics for Epidemiology* (London: Chapman & Hall/CRC, 2004); M. Macpherson, B. Naughton, A. Hsu, and J. Mountain, "Estimating Genotype-Specific Incidence for One or Several Loci," 23andMe, 2007; N. Risch, "Linkage Strategies for Genetically Complex Traits. I. Multilocus models," *American Journal of Human Genetics* 46 (2) (1990): 222–28.

46 *the findings of a 2014 study that investigated the risk-calculation methods:* R. R. J. Kalf et al., "Variations in Predicted Risks in Personal Genome Testing for Common Complex Diseases," *Genetics in Medicine* 16 (1) (2014): 85–91, https://doi.org/10.1038/gim.2013.80.

47 *BMI was first cooked up in 1835 by Belgian Adolphe Quetelet:* L. A. J. Quetelet, "A Treatise on Man and the Development of His Faculties," *Obesity Research* 2 (1) (1994): 72–85, https://doi.org/10.1002/j.1550-8528.1994 .tb00047.x.

48 *Ancel Keys . . . undertook a study:* A. Keys et al., "Indices of Relative Weight and Obesity," *Journal of Chronic Diseases* 25 (6–7) (1972): 329–43, https:// doi.org/10.1016/0021-9681(72)90027-6.

48 *between 15 and 35 percent of men with non-obese BMIs would be reclassified as obese:* A. J. Tomiyama, J. M. Hunger, J. Nguyen-Cuu, and C. Wells, "Misclassification of Cardiometabolic Health When Using Body Mass Index

Categories in NHANES, 2005–2012," *International Journal of Obesity* 40 (5) (2016): 883–86, https://doi.org/10.1038/ijo.2016.17.

49 *These incorrect classifications have implications*: R. L. McCrea, Y. G. Berger, and M. B. King, "Body Mass Index and Common Mental Disorders: Exploring the Shape of the Association and Its Moderation by Age, Gender and Education," *International Journal of Obesity* 36 (3) (2012): 414–21, https://doi.org/10.1038/ijo.2011.65.

53 *approximately 85 percent of automated warnings in ICUs are false alarms*: S. Sendelbach and M. Funk, "Alarm Fatigue: A Patient Safety Concern," *AACN Advanced Critical Care* 24 (4) (2013): 378–86, quiz on 387–88, https://doi.org/10.1097/NCI.0b013e3182a903f9; S. T. Lawless, "Crying Wolf: False Alarms in a Pediatric Intensive Care Unit," *Critical Care Medicine* 22 (6) (1994): 981–85.

56 *median filtering is beginning to be used in our ICU monitors*: A. Mäkivirta, E. Koski, A. Kari, and T. Sukuvaara, "The Median Filter as a Preprocessor for a Patient Monitor Limit Alarm System in Intensive Care," *Computer Methods and Programs in Biomedicine* 34 (2–3) (1991): 139–44, https://doi.org/10.1016/0169-2607(91)90039-V.

56 *reduce the occurrence of false alarms in ICU monitors by as much as 60 percent*: M. Imhoff, S. Kuhls, U. Gather, and R. Fried, "Smart Alarms from Medical Devices in the OR and ICU," *Best Practice & Research Clinical Anaesthesiology* 23 (1) (2009): 39–50, https://doi.org/10.1016/J.BPA.2008.07.008.

58 *the test will pick this up roughly nine times out of ten*: S. Hofvind, B. M. Geller, J. Skelly, and P. M. Vacek, "Sensitivity and Specificity of Mammographic Screening as Practised in Vermont and Norway," *British Journal of Radiology* 85 (1020) (2012): e1226–32, https://doi.org/10.1259/bjr/15168178.

59 *In 2007, 160 gynecologists were given the following information*: G. Gigerenzer et al., "Helping Doctors and Patients Make Sense of Health Statistics," *Psychological Science in the Public Interest* 8 (2) (2007): 53–96, https://doi.org/10.1111/j.1539-6053.2008.00033.x.

62 *Writing in the* British Medical Journal, *Muir Gray*: J. A. M. Gray, J. Patnick, and R. G. Blanks, "Maximising Benefit and Minimising Harm of Screening," *BMJ, Clinical Research Edition* 336 (7642) (2008): 480–83, https://doi.org/10.1136/bmj.39470.643218.94.

63 *one thousand adults in Germany were asked*: Gigerenzer et al., "Helping Doctors and Patients," 53–96.

64 *ELISA had reported false-positive rates of around 0.3 percent*: J. K. Cornett and T. J. Kirn, "Laboratory Diagnosis of HIV in Adults: A Review of Current Methods," *Clinical Infectious Diseases* 57 (5) (2013): 712–18, https://doi.org/10.1093/cid/cit281.

67 *a global team of researchers developed a blood test*: D. Bougard et al., "Detection of Prions in the Plasma of Presymptomatic and Symptomatic Patients

with Variant Creutzfeldt-Jakob Disease," *Science Translational Medicine* 8 (370) (2016): 370ra182, https://doi.org/10.1126/scitranslmed.aag1257.

70 *a woman miscarried after she was sanctioned*: C. S. Sigel and D. G. Grenache, "Detection of Unexpected Isoforms of Human Chorionic Gonadotropin by Qualitative Tests," *Clinical Chemistry* 53 (5) (2007): 989–90, https://doi .org/10.1373/clinchem.2007.085399.

70 *Another woman's ectopic pregnancy was missed by urine tests*: A. Daniilidis et al., "A Unique Case of Ruptured Ectopic Pregnancy in a Patient with Negative Pregnancy Test—a Case Report and Brief Review of the Literature," *Hippokratia* 18 (3) (2014): 282–84.

Chapter 3: The Laws of Mathematics

77 *"must have been done carefully on purpose"*: L. Schneps and C. Colmez, *Math on Trial: How Numbers Get Used and Abused in the Courtroom* (New York: Basic Books, 2013).

78 *Poincaré debunked the aberrant handwriting analysis*: Jean Mawhin. "Henri Poincaré: A Life in the Service of Science," *Notices of the American Mathematical Society* 52 (9) (2005): 1036–44.

79 *Japanese criminal justice system, for example, has a conviction rate of 99.9 percent*: J. M. Ramseyer and E. B Rasmusen, "Why Is the Japanese Conviction Rate So High?," *Journal of Legal Studies* 30 (1) (2001): 53–88, https://doi .org/10.1086/468111.

81 *"One sudden infant death is a tragedy, two is suspicious"*: R. Meadow, ed., *ABC of Child Abuse* (London: British Medical Journal Publishing Group, 1989).

82 *The prevalence of autism in the United States is roughly 1 per 100*: C. Rice, "Prevalence of Autism Spectrum Disorders," *Morbidity and Mortality Weekly Report* 58 (10) (2009): 1–20, http://www.ncbi.nlm.nih.gov/pubmed/20023608.

82 *Only 1 in 5 of those on the autistic spectrum are female*: S. Ehlers and C. Gillberg, "The Epidemiology of Asperger Syndrome," *Journal of Child Psychology and Psychiatry* 34 (8) (1993): 1327–50, https://doi.org/10.1111/j.1469-7610.1993 .tb02094.x.

83 *report on SIDS for which he had been asked to write the preface*: P. J. Fleming, P. S. P. Blair, C. Bacon, and P. J. Berry, *Sudden Unexpected Deaths in Infancy: The CESDI SUDI Studies, 1993–1996* (London: Stationery Office, 2000); C. E. A. Leach et al., "Epidemiology of SIDS and Explained Sudden Infant Deaths," *Pediatrics* 104 (4) (1999).

84 *researchers at the University of Manchester also identified markers*: A. M. Summers et al., "Association of IL-10 Genotype with Sudden Infant Death Syndrome," *Human Immunology* 61 (12) (2000): 1270–73, https://doi .org/10.1016/S0198-8859(00)00183-X.

84 *Many more genetic risk factors have since been identified:* C. A. Brownstein, A. Poduri, R. D. Goldstein, and I. A. Holm, "The Genetics of Sudden Infant Death Syndrome," in *SIDS: Sudden Infant and Early Childhood Death: The Past, the Present and the Future,* ed. Jodhie R. Duncan and Roger W. Byard (South Australia: University of Adelaide, 2018); M. Dashash et al., "Association of Sudden Infant Death Syndrome with VEGF and IL-6 Gene Polymorphisms," *Human Immunology* 67 (8) (2006): 627–33, https://doi.org/10.1016/J.HUMIMM.2006.05.002.

89 *measuring the health of the economy:* Y. Z. Ma, "Simpson's Paradox in GDP and Per Capita GDP Growths," *Empirical Economics* 49 (4) (2015): 1301–15, https://doi.org/10.1007/s00181-015-0921-3.

89 *understanding voter profiles:* H. Nurmi, "Voting Paradoxes and Referenda," *Social Choice and Welfare* 15 (3) (1998): 333–50, https://doi.org/10.1007/s003550050109.

89 *in drug development:* N. S. Abramson, S. F. Kelsey, P. Safar, and K. Sutton-Tyrrell, "Simpson's Paradox and Clinical Trials: What You Find Is Not Necessarily What You Prove," *Annals of Emergency Medicine* 21 (12) (1992): 1480–82, https://doi.org/10.1016/S0196-0644(05)80066-6.

92 *smoking during pregnancy was providing some protection:* J. Yerushalmy, "The Relationship of Parents' Cigarette Smoking to Outcome of Pregnancy—Implications as to the Problem of Inferring Causation from Observed Associations," *American Journal of Epidemiology* 93 (6) (1971): 443–56, https://doi.org/10.1093/oxfordjournals.aje.a121278.

92 *it was nothing of the sort:* A. J. Wilcox, "On the Importance—and the Unimportance—of Birthweight," *International Journal of Epidemiology* 30 (6) (2001): 1233–41, https://doi.org/10.1093/ije/30.6.1233.

96 *Double infant murder has been calculated:* A. P. Dawid, "Bayes's Theorem and Weighing Evidence by Juries," in *Bayes's Theorem,* ed. Richard Swinburne (London: British Academy, 2005), https://doi.org/10.5871/bacad/9780197263419.003.0004; R. Hill, "Multiple Sudden Infant Deaths—Coincidence or Beyond Coincidence?," *Paediatric and Perinatal Epidemiology* 18 (5) (2004): 320–26, https://doi.org/10.1111/j.1365-3016.2004.00560.x.

102 *suggest that Judge Hellmann was wrong:* Schneps and Colmez, *Math on Trial.*

105 *the use of cranberries to treat urinary tract infections:* R. G. Jepson, G. Williams, and J. C. Craig, "Cranberries for Preventing Urinary Tract Infections," *Cochrane Database of Systematic Reviews* (10) (2012), https://doi.org/10.1002/14651858.CD001321.pub5.

105 *the use of vitamin C for preventing the common cold:* H. Hemilä, E. Chalker, and B. Douglas, "Vitamin C for Preventing and Treating the Common Cold," *Cochrane Database of Systematic Reviews* (3) (2007), https://doi.org/10.1002/14651858.CD000980.pub3.

Chapter 4: Don't Believe the Truth

111 *Truthfulness and accuracy are near the top*: American Society of News Editors, "ASNE Statement of Principles," 2019, https://www.asne.org/content. asp?pl=24&sl=171&contentid=171; International Federation of Journalists, "Principles on Conduct of Journalism — IFJ," 2019, https://www.ifj.org/who/ rules-and-policy/principles-on-conduct-of-journalism.html; Associated Press Media Editors, "Statement of Ethical Principles — APME," 2019, https://www .apme.com/page/EthicsStatement?&hhsearchterms=%22ethics%22; Society of Professional Journalists, "SPJ Code of Ethics," 2019, https://www.spj.org/ ethicscode.asp.

118 *Based on this study*: K. Troyer, T. Gilboy, and B. Koeneman, "A Nine STR Locus Match between Two Apparently Unrelated Individuals Using Amp-FlSTR® Profiler Plus and Cofiler," *Genetic Identity Conference Proceedings, 12th International Symposium on Human Identification*, 2001, https://www .promega.ee/~/media/files/resources/conference proceedings/ishi 12/poster abstracts/troyer.pdf.

118 *If 122 matches had turned up in a database*: J. Curran, "Are DNA Profiles as Rare as We Think? Or Can We Trust DNA Statistics?," *Significance* 7 (2) (2010): 62–66, https://doi.org/10.1111/j.1740-9713.2010.00420.x.

122 *Federal Trade Commission (FTC) wrote to L'Oréal*: E. Ramirez et al., "In the Matter of L'Oréal USA Inc., a Corporation. Docket No. C.," 2014, https:// www.ftc.gov/system/files/documents/cases/140627lorealcmpt.pdf.

123 *they had predicted Roosevelt's victory margin to within a percentage point*: P. Squire, "Why the 1936 *Literary Digest* Poll Failed," *Public Opinion Quarterly* 52 (1) (1988): 125, https://doi.org/10.1086/269085.

123 *In August, they sent out straw polls*: J. L. Simon, *The Art of Empirical Investigation* (Piscataway, NJ: Transaction Publishers, 2003).

123 *"Landon, 1,293,669; Roosevelt, 972,897"*: "Landon, 1,293,669; Roosevelt, 972,897: Final Returns in 'The Digest's' Poll of Ten Million Voters," *Literary Digest* 122 (1936): 5–6.

124 Fortune *magazine predicted the margin of Roosevelt's victory*: H. Cantril, "How Accurate Were the Polls?," *Public Opinion Quarterly* 1 (1) (1937): 97, https://doi.org/10.1086/265040; D. Lusinchi, " 'President' Landon and the 1936 *Literary Digest* Poll," *Social Science History* 36 (1) (2012): 23–54, https://doi.org/10.1017/S014555320001035X.

124 *magazine's demise less than two years later*: Squire, "Why the 1936 *Literary Digest* Poll Failed," 125.

126 A *mathematically oriented blog post*: R. Old, "Rod Liddle said, 'Do the math,' So I did," polarizingthevacuum, 2016, https://polarizingthevacuum.wordpress .com/2016/09/08/rod-liddle-said-do-the-math-so-i-did/#comments.

126 *According to FBI statistics*: Federal Bureau of Investigation, "Crime in the

United States: FBI—Expanded Homicide Data Table 6," 2015, https://ucr
.fbi.gov/crime-in-the-u.s/2015/crime-in-the-u.s.-2015/tables/expanded_homi-
cide_data_table_6_murder_race_and_sex_of_vicitm_by_race_and_sex_of_
offender_2015.xls.

126 *black people comprised just 12.6 percent of the US population:* US Census
Bureau, "American FactFinder—Results," 2015, https://factfinder.census
.gov/bkmk/table/1.0/en/ACS/15_5YR/DP05/0100000US.

127 *FBI was found to be recording fewer than half of all killings by police:* J. Swaine,
O. Laughland, J. Lartey, and C. McCarthy, "The Counted: People Killed
by Police in the US," 2016, https://www.theguardian.com/us-news/series/
counted-us-police-killings.

127 *"embarrassing and ridiculous" that the* Guardian *had better data:* M. Tran,
"FBI Chief: 'Unacceptable' That *Guardian* Has Better Data on Police
Violence," *Guardian,* October 8, 2015, https://www.theguardian.com/
us-news/2015/oct/08/fbi-chief-says-ridiculous-guardian-washington-post
-better-information-police-shootings.

128 *only 635,781 full-time "law enforcement officers":* Federal Bureau of Investigation, "Crime in the United States: Full-Time Law Enforcement Employees,"
2015, https://ucr.fbi.gov/crime-in-the-u.s/2015/crime-in-the-u.s.-2015/tables/
table-74.

130 *effect of consuming fifty grams of processed meat per day:* World Cancer
Research Fund and American Institute for Cancer Research, *Second Expert
Report | World Cancer Research Fund International,* 2007, http://discovery
.ucl.ac.uk/4841/1/4841.pdf.

132 *reported in the journal* Nature Genetics: C. Newton-Cheh et al., "Association of Common Variants in NPPA and NPPB with Circulating Natriuretic
Peptides and Blood Pressure," *Nature Genetics* 41 (3) (2009): 348–53, https://
doi.org/10.1038/ng.328.

133 *In one study from 2010:* R. Garcia-Retamero and M. Galesic, "How to Reduce
the Effect of Framing on Messages about Health," *Journal of General Internal
Medicine* 25 (12) (2010): 1323–29, https://doi.org/10.1007/s11606-010-1484-9.

133 *mismatched framing and was found to occur in roughly a third of the articles:*
A. Sedrakyan and C. Shih, "Improving Depiction of Benefits and Harms,"
Medical Care 45 (10 suppl. 2) (2007): S23–S28, https://doi.org/10.1097/
MLR.0b013e3180642f69.

133 *results of a recent clinical trial:* B. Fisher et al., "Tamoxifen for Prevention of
Breast Cancer: Report of the National Surgical Adjuvant Breast and Bowel
Project P-1 Study," *Journal of the National Cancer Institute* 90 (18) (1998):
1371–88, https://doi.org/10.1093/jnci/90.18.1371.

134 *Using percentages instead of decimals to highlight perceived benefits:* G. Pas-
serini, L. Macchi, and M. Bagassi, "A Methodological Approach to Ratio
Bias," *Judgment and Decision Making* 7 (5) (2012).

134 *Our susceptibility to ratio bias has been confirmed:* V. Denes-Raj and S. Epstein, "Conflict between Intuitive and Rational Processing: When People Behave Against Their Better Judgment," *Journal of Personality and Social Psychology* 66 (5) (1994): 819–29, https://doi.org/10.1037/0022-3514.66.5.819.

137 *In one study conducted in 1987:* H. C. Faigel, "The Effect of Beta Blockade on Stress-Induced Cognitive Dysfunction in Adolescents," *Clinical Pediatrics* 30 (7) (1991): 441–45, https://doi.org/10.1177/000992289103000706.

138 *majority of the so-called placebo effect is actually a result of regression to the mean:* A. Hróbjartsson and P. C. Gøtzsche, "Placebo Interventions for All Clinical Conditions," *Cochrane Database of Systematic Reviews* (1) (2010), https://doi.org/10.1002/14651858.CD003974.pub3.

141 *first studies comparing crime rates:* J. R. Lott, *More Guns, Less Crime: Understanding Crime and Gun Control Laws,* 2nd ed. (Chicago: University of Chicago Press, 2000); J. R. Lott Jr. and D. B. Mustard, "Crime, Deterrence, and Right-to-Carry Concealed Handguns, *Journal of Legal Studies* 26 (1) (1997): 1–68, https://doi.org/10.1086/467988; F. Plassmann and T. N. Tideman, "Does the Right to Carry Concealed Handguns Deter Countable Crimes? Only a Count Analysis Can Say," *Journal of Law and Economics* 44 (S2) (2001): 771–98, https://doi.org/10.1086/323311; W. A. Bartley and M. A. Cohen "The Effect of Concealed Weapons Laws: An Extreme Bound Analysis," *Economic Inquiry* 36 (2) (1998): 258–65, https://doi.org/10.1111/j.1465-7295.1998.tb01711.x; C. E. Moody, "Testing for the Effects of Concealed Weapons Laws: Specification Errors and Robustness," *Journal of Law and Economics* 44 (S2) (2001): 799–813, https://doi.org/10.1086/323313.

141 *policing, rising numbers of incarcerations, and the receding crack cocaine epidemic:* S. D. Levitt, "Understanding Why Crime Fell in the 1990s: Four Factors That Explain the Decline and Six That Do Not," *Journal of Economic Perspectives* 18 (1) (2004): 163–90, https://doi.org/10.1257/089533004773563485.

141 *"gives no support to the hypothesis that shall-issue laws":* P. Grambsch, "Regression to the Mean, Murder Rates, and Shall-Issue Laws," *American Statistician* 62 (4) (2008): 289–95, https://doi.org/10.1198/000313008X362446.

Chapter 5: Wrong Place, Wrong Time

161 *simple rounding error in a German election in 1992:* D. Weber-Wulff, "Rounding Error Changes Parliament Makeup," *Risks Digest* 13 (37) (1992).

161 *newly created Vancouver Stock Exchange index plummeted continuously:* B. D. McCullough and H. D. Vinod, "The Numerical Reliability of Econometric Software," *Journal of Economic Literature* 37 (2) (1999): 633–65, https://doi.org/10.1257/jel.37.2.633.

163 *United States remains the last industrial nation to use imperial units:* Technically, US customary units are slightly different from their close relatives of

the British imperial system. The differences, however, are not important for the purposes of this book, so we will refer to both measurement systems as "imperial."

172 *missile alert was assumed to be a false alarm:* H. Wolpe, *Patriot Missile Defense: Software Problem Led to System Failure at Dhahran, Saudi Arabia* (Washington, DC: US General Accounting Office, 1992), https://www.gao.gov/products/IMTEC-92-26.

Chapter 6: Relentless Optimization

178 *In 2000, the Clay Mathematics Institute announced:* A. M. Jaffe, "The Millennium Grand Challenge in Mathematics," *Notices of the AMS* 53 (6) (2006).

179 *Russian mathematician Grigori Perelman shared three dense mathematical papers:* G. Perelman, "The Entropy Formula for the Ricci Flow and Its Geometric Applications," 2002, http://arxiv.org/abs/math/0211159; "Finite Extinction Time for the Solutions to the Ricci Flow on Certain Three-Manifolds," 2003, http://arxiv.org/abs/math/0307245; and "Ricci Flow with Surgery on Three-Manifolds," 2003, http://arxiv.org/abs/math/0303109.

185 *probably worth consulting Cook's algorithm first:* W. Cook, *In Pursuit of the Traveling Salesman: Mathematics at the Limits of Computation* (Princeton, NJ: Princeton University Press, 2012).

187 *Dijkstra's algorithm—finds the solution:* E. W. Dijkstra, "A Note on Two Problems in Connexion with Graphs," *Numerische Mathematik* 1 (1) (1959): 269–71.

195 *mathematical shorthand for a number known as Euler's number:* Euler's number first appeared in the seventeenth century, when Swiss mathematician Jacob Bernoulli (uncle of the early mathematical biologist Daniel Bernoulli, whose epidemiological exploits are relayed in chapter 7) was investigating compound interest. In chapter 1 we encountered compound interest, which means that interest is paid into the account so that it can accrue interest itself. Bernoulli wanted to know how the amount of interest accrued at the end of a year depends on how often the interest is compounded.

Imagine, for simplicity, that the bank pays a special rate of 100 percent a year on an initial investment of £1. Interest is added to the account at the end of each fixed period, and interest can then be paid on that interest in the next period. What happens if the bank decides to pay interest only once a year? At the end of the year, we receive £1 in interest, but there is no time left to accrue further interest on the interest, so we are left with £2. Alternatively, if the bank decides to pay us every six months, then after half a year the bank calculates the interest owed using half the yearly rate (i.e., 50 percent) leaving us with £1.50 in the account. The same procedure is repeated at the end of the year, giving 50 percent interest on the £1.50 in the account, and leaving a total of £2.25 at the end of the year.

By compounding more often, the money in the account by the end of the year increases. Compounding quarterly, for example, gives £2.44; monthly compounding yields £2.61. Bernoulli showed that by using continuous compounding (i.e., calculating and accruing interest infinitely often, but with an infinitely small rate), the amount of money at year end would peak at approximately £2.72. To be more precise, we would have precisely e (Euler's number) pounds at the end of the year.

196 *came to the attention of mathematicians as the "hiring problem"*: T. S. Ferguson, "Who Solved the Secretary Problem?," *Statistical Science* 4 (3) (1989): 282–89, https://doi.org/10.1214/ss/1177012493; J. P. Gilbert and F. Mosteller, "Recognizing the Maximum of a Sequence," *Journal of the American Statistical Association* 61 (313) (1966): 35, https://doi.org/10.2307/2283044.

Chapter 7: Susceptible, Infective, Removed

210 *In the year 2000, measles was officially declared eliminated*: A. P. Fiebelkorn et al., "A Comparison of Postelimination Measles Epidemiology in the United States, 2009–2014 versus 2001–2008," *Journal of the Pediatric Infectious Diseases Society* 6 (1) (2017): 40–48, https://doi.org/10.1093/jpids/piv080.

212 *Jenner carried out a pioneering experiment into disease prevention*: E. Jenner, *An inquiry into the causes and effects of the variolae vaccinae, a disease discovered in some of the western counties of England, particularly Gloucestershire, and known by the name of the cow pox*, ed. S. Low (London, 1798).

214 *This method was not supplanted by a less invasive alternative for over 170 years*: J. Booth, "A Short History of Blood Pressure Measurement," *Proceedings of the Royal Society of Medicine* 70 (11) (1977): 793–99.

214 *Bernoulli suggested an equation to describe the proportion of people*: D. Bernoulli and S. Blower, "An Attempt at a New Analysis of the Mortality Caused by Smallpox and of the Advantages of Inoculation to Prevent It," *Reviews in Medical Virology* 14 (5) (2004): 275–88, https://doi.org/10.1002/rmv.443.

215 *poor sanitation and crowded living environments in colonial India*: J. N. Hays, *Epidemics and Pandemics: Their Impacts on Human History* (Santa Barbara, CA: ABC-CLIO, 2005); S. Watts, "British Development Policies and Malaria in India, 1897–c. 1929," *Past & Present* 165 (1) (1999): 141–81, https://doi.org/10.1093/past/165.1.141; M. Harrison, "'Hot Beds of Disease': Malaria and Civilization in Nineteenth-Century British India," *Parassitologia* 40 (1–2) (1998): 11–18, http://www.ncbi.nlm.nih.gov/pubmed/9653727; M. U. Mushtaq, "Public Health in British India: A Brief Account of the History of Medical Services and Disease Prevention in Colonial India," *Indian Journal of Community Medicine: Official Publication of Indian Association of Preventive & Social Medicine* 34 (1) (2009): 6–14, https://doi.org/10.4103/0970-0218.45369.

215 *No one is entirely sure how the disease reached Bombay:* W. J. Simpson, *A Treatise on Plague Dealing with the Historical, Epidemiological, Clinical, Therapeutic and Preventive Aspects of the Disease* (Cambridge, UK: Cambridge University Press, 2010), https://doi.org/10.1017/CBO9780511710773.

216 *they conducted the single most influential study:* W. O. Kermack and A. G. McKendrick, "A Contribution to the Mathematical Theory of Epidemics," *Proceedings of the Royal Society A: Mathematical, Physical and Engineering Sciences* 115 (772) (1927): 700–721, https://doi.org/10.1098/rspa.1927.0118.

217 *over a thousand outbreaks of the vomiting bug, norovirus, were linked:* A. J. Hall et al., "Vital Signs: Foodborne Norovirus Outbreaks—United States, 2009–2012," *Morbidity and Mortality Weekly Report* 63 (22) (2014): 491–95.

219 *outbreaks die out from of a lack of infective people:* J. D. Murray, *Mathematical Biology I: An Introduction* (New York: Springer, 2002).

221 *Over 60 percent of all cervical cancers are caused by two strains:* F. X. Bosch et al., "Prevalence of Human Papillomavirus in Cervical Cancer: A Worldwide Perspective. International Biological Study on Cervical Cancer (IBSCC) Study Group," *Journal of the National Cancer Institute* 87 (11) (1995): 796–802.

221 *HPV is the most frequent sexually transmitted disease in the world:* N. Gavillon et al., "Papillomavirus Humain (HPV): Comment Ai-Je Attrapé Ça?," *Gynécologie Obstétrique & Fertilité* 38 (3) (2010): 199–204, https://doi .org/10.1016/J.GYOBFE.2010.01.003.

221 *Studies undertaken in the UK around the time of the vaccine's deployment:* M. Jit, Y. H. Choi, and W. J. Edmunds, "Economic Evaluation of Human Papillomavirus Vaccination in the United Kingdom," *BMJ, Clinical Research Edition* 337 (2008): a769, https://doi.org/10.1136/bmj.a769.

221 *Related studies in other countries, considering mathematical models:* I. Zechmeister et al., "Cost-Effectiveness Analysis of Human Papillomavirus-Vaccination Programs to Prevent Cervical Cancer in Austria," *Vaccine* 27 (37) (2009): 5133–41, https://doi.org/10.1016/J.VACCINE.2009.06.039.

222 *the strains of HPV guarded against by the vaccine can also cause a range:* M. Kohli et al., "Estimating the Long-Term Impact of a Prophylactic Human Papillomavirus 16/18 Vaccine on the Burden of Cervical Cancer in the UK," *British Journal of Cancer* 96 (1) (2007): 143–50, https://doi.org/10.1038/ sj.bjc.6603501; S. L. Kulasingam et al., "Adding a Quadrivalent Human Papillomavirus Vaccine to the UK Cervical Cancer Screening Programme: A Cost-Effectiveness Analysis," *Cost Effectiveness and Resource Allocation* 6 (1) (2008): 4, https://doi.org/10.1186/1478-7547-6-4; E. Dasbach, R. Insinga, and E. Elbasha, "The Epidemiological and Economic Impact of a Quadrivalent Human Papillomavirus Vaccine (6/11/16/18) in the UK," *BJOG: An International Journal of Obstetrics & Gynaecology* 115 (8) (2008): 947–56, https://doi.org/10.1111/j.1471-0528.2008.01743.x.

222 *As well as causing cervical cancer, HPV types 16 and 18:* S. Hibbitts, "Should Boys Receive the Human Papillomavirus Vaccine? Yes," *BMJ* 339 (2009): b4928, https://doi.org/10.1136/BMJ.B4928; D. M. Parkin and F. Bray, "Chapter 2: The Burden of HPV-Related Cancers," *Vaccine* 24 (2006): S11–S25, https://doi.org/10.1016/J.VACCINE.2006.05.111; M. Watson et al., "Using Population-Based Cancer Registry Data to Assess the Burden of Human Papillomavirus-Associated Cancers in the United States: Overview of Methods," *Cancer* 113 (S10) (2008): 2841–54, https://doi.org/10.1002/cncr.23758.

222 *In both the United States and the United Kingdom, the majority of cancers caused by HPV are not cervical:* Hibbitts, "Should Boys Receive," b4928; ICO/IARC Information Centre on HPV and Cancer, "United Kingdom Human Papillomavirus and Related Cancers, Fact Sheet 2018"; Watson et al., "Using Population-Based Cancer Registry Data," 2841–54.

222 *HPV types 6 and 11 also cause nine out of ten cases of anogenital warts:* V. R. Yanofsky, R. V. Patel, and G. Goldenberg, "Genital Warts: A Comprehensive Review," *Journal of Clinical and Aesthetic Dermatology* 5 (6) (2012): 25–36.

222 *60 percent of the health-care costs associated with all noncervical HPV infections:* D. Hu and S. Goldie, "The Economic Burden of Noncervical Human Papillomavirus Disease in the United States," *American Journal of Obstetrics and Gynecology* 198 (5) (2008): 500.e1–500.e7, https://doi.org/10.1016/J.AJOG.2008.03.064.

223 *Models based on sexual networks including homosexual relationships:* J. Gómez-Gardeñes, V. Latora, Y. Moreno, and E. Profumo, "Spreading of Sexually Transmitted Diseases in Heterosexual Populations," *Proceedings of the National Academy of Sciences of the United States of America* 105 (5) (2008): 1399–1404, https://doi.org/10.1073/pnas.0707332105.

223 *The prevalence of HPV in men who have sex with men:* M. M. Blas et al., "HPV Prevalence in Multiple Anatomical Sites among Men Who Have Sex with Men in Peru," *PLoS ONE* 10 (10) (2015): e0139524, https://doi.org/10.1371/journal.pone.0139524; G. McQuillan et al., "Prevalence of HPV in Adults Aged 18–69: United States, 2011–2014," *NCHS Data Brief* (280) (2017): 1–8, http://www.ncbi.nlm.nih.gov/pubmed/28463105.

223 *incidence rate of anal cancer in this group is over fifteen times higher:* G. D'Souza et al., "Incidence and Epidemiology of Anal Cancer in the Multicenter AIDS Cohort Study," *Journal of Acquired Immune Deficiency Syndromes* 48 (4) (2008): 491–99, https://doi.org/10.1097/QAI.0b013e31817aebfe; L. G. Johnson et al., "Anal Cancer Incidence and Survival: The Surveillance, Epidemiology, and End Results Experience, 1973–2000," *Cancer* 101 (2) (2004): 281–88, https://doi.org/10.1002/cncr.20364; J. R. Qualters, N. C. Lee, R. A. Smith, and R. E. Aubert, "Breast and Cervical Cancer Surveillance, United States, 1973–1987," *Morbidity and Mortality Weekly Report: Surveillance Summaries*, Centers for Disease Control and Prevention (CDC),

1987; US Cancer Statistics Working Group, "U.S. Cancer Statistics Data Visualizations Tool," based on November 2017 submission data (1999–2015), US Department of Health and Human Services, Centers for Disease Control and Prevention and National Cancer Institute, June 2018, www.cdc.gov /cancer/dataviz; A. M. Noone et al., eds., "SEER Cancer Statistics Review, 1975–2015," external based on November 2017 SEER data submission, National Cancer Institute (Bethesda, MD), April 2018, https://seer.cancer .gov/csr/1975_2015/; P. V. Chin-Hong et al., "Age-Specific Prevalence of Anal Human Papillomavirus Infection in HIV—Negative Sexually Active Men Who Have Sex with Men: The EXPLORE Study," *Journal of Infectious Diseases* 190 (12) (2004): 2070–76, https://doi.org/10.1086/425906.

223 *advice based on a new cost-effectiveness study recommended that all boys:* M. Brisson et al., "Population-Level Impact, Herd Immunity, and Elimination after Human Papillomavirus Vaccination: A Systematic Review and Meta-analysis of Predictions from Transmission-Dynamic Models," *Lancet: Public Health* 1 (1) (2016): e8–e17, https://doi.org/10.1016/S2468-2667(16)30001-9; M. J. Keeling, K. A. Broadfoot, and S. Datta, "The Impact of Current Infection Levels on the Cost-Benefit of Vaccination," *Epidemics* 21 (2017): 56–62, https://doi.org/10.1016/J.EPIDEM.2017.06.004; Joint Committee on Vaccination and Immunisation, "Statement on HPV Vaccination," 2018, https://www.gov.uk/government/publications/jcvi-statement -extending-the-hpv-vaccination-programme-conclusions; Joint Committee on Vaccination and Immunisation, "Interim Statement on Extending the HPV Vaccination Programme," 2018, https://www.gov.uk/government/publications/ jcvi-statement-extending-the-hpv-vaccination-programme.

227 *simple mathematical model incorporating an incubation period:* D. Mabey, S. Flasche, and W. J. Edmunds, "Airport Screening for Ebola," *BMJ, Clinical Research Edition* 349 (2014): g6202, https://doi.org/10.1136/bmj.g6202.

232 *simple applications of mathematical modeling can suggest:* C. Castillo-Chavez, C. W. Castillo-Garsow, and A.-A. Yakubu, "Mathematical Models of Isolation and Quarantine," *Journal of the American Medical Association* 290 (21) (2003): 2876–77, https://doi.org/10.1001/jama.290.21.2876.

233 *a quarantining strategy succeeds depends on the* timing *of peak infectiousness:* T. Day et al., "When Is Quarantine a Useful Control Strategy for Emerging Infectious Diseases?," *American Journal of Epidemiology* 163 (5) (2006): 479–85, https://doi.org/10.1093/aje/kwj056; C. M. Peak et al., "Comparing Nonpharmaceutical Interventions for Containing Emerging Epidemics," *Proceedings of the National Academy of Sciences of the United States of America* 114 (15) (2017): 4023–28, https://doi.org/10.1073/pnas.1616438114.

233 *mathematical study concluded that approximately 22 percent:* F. B. Agusto, M. I. Teboh-Ewungkem, and A. B. Gumel, "Mathematical Assessment of the Effect of Traditional Beliefs and Customs on the Transmission Dynamics

of the 2014 Ebola Outbreaks," *BMC Medicine* 13 (1) (2015): 96, https://doi .org/10.1186/s12916-015-0318-3.

236 *A study that modeled the spread of the 2015 Disneyland measles outbreak:* M. S. Majumder et al., "Substandard Vaccination Compliance and the 2015 Measles Outbreak," *JAMA Pediatrics* 169 (5) (2015): 494, https://doi.org/10.1001/ jamapediatrics.2015.0384.

237 *somber five-page publication in the well-respected medical journal the* Lancet: A. Wakefield et al., "RETRACTED: Ileal-Lymphoid-Nodular Hyperplasia, Non-specific Colitis, and Pervasive Developmental Disorder in Children," *Lancet* 351 (9103) (1998): 637–41, https://doi.org/10.1016/S0140-6736(97)11096-0.

240 *World Health Organization figures show that vaccines prevent millions of deaths:* World Health Organization: Strategic Advisory Group of Experts on Immunization, *SAGE DoV GVAP Assessment Report 2018*, World Health Organization, 2018, https://www.who.int/immunization/global_vaccine_ action_plan/sage_assessment_reports/en/.

INDEX

Page numbers in *italics* refer to illustrations

ABOUT THE AUTHOR

Kit Yates is a senior lecturer in the Department of Mathematical Sciences and codirector of the Centre for Mathematical Biology at the University of Bath. He completed his PhD in mathematics at the University of Oxford in 2011. His research into mathematical biology has been covered by the BBC, the *Guardian*, the *Telegraph*, the *Daily Mail*, *Scientific American*, and Reuters, among other media outlets. *The Math of Life and Death* is his first book.